An Introduction to the Invertebrates

So much has to be crammed into today's biology courses that basic information on animal groups and their evolutionary origins is often left out. This is particularly true of invertebrates, where most of the texts available are too large for the typical short introductory course. This volume fills the gap. *An Introduction to the Invertebrates* is a short guide to the invertebrate phyla, looking at their diverse forms, functions and evolutionary relationships. This book first introduces evolution and molecular methods of tracing it, then considers the distinctive body plan of each invertebrate phylum, showing what has evolved, how the animals live, and how they develop. Boxes introduce physiological mechanisms and development. The final chapter combines fossil and molecular evidence to present an up-to-date view of evolutionary history, giving a more certain definition of the relationships between invertebrates. This user-friendly and well-illustrated introduction will be invaluable for all those studying invertebrates.

JANET MOORE is former Director of Studies in Biological Sciences at New Hall Cambridge. Her research career has focused on land and freshwater nemertines. Now officially retired, she nevertheless continues to teach zoology to undergraduate students at Cambridge.

The **Studies in Biology** series is published in association with the Institute of Biology (London, UK). The series provides short, affordable and very readable textbooks aimed primarily at undergraduate biology students. Each book offers either an introduction to a broad area of biology (e.g. *Introductory Microbiology*), or a more in-depth treatment of a particular system or specific topic (e.g. *Photosynthesis*). All of the subjects and systems covered are selected on the basis that all undergraduate students will study them at some point during their biology degree courses.

Titles available in this series

An Introduction to the Invertebrates

Janet Moore

Emeritus Fellow of New Hall, Cambridge, and
Researcher, Department of Zoology, Cambridge

Illustrations by Raith Overhill

CAMBRIDGE
UNIVERSITY PRESS

PUBLISHED BY THE PRESS SYNDICATE OF THE UNIVERSITY OF CAMBRIDGE
The Pitt Building, Trumpington Street, Cambridge, United Kingdom

CAMBRIDGE UNIVERSITY PRESS
The Edinburgh Building, Cambridge CB2 2RU, UK
40 West 20th Street, New York, NY 10011-4211, USA
10 Stamford Road, Oakleigh, VIC 3166, Australia
Ruiz de Alarcón 13, 28014 Madrid, Spain
Dock House, The Waterfront, Cape Town 8001, South Africa

http://www.cambridge.org

First published 2001

Printed in the United Kingdom at the University Press, Cambridge

Typeface Monotype Garamond 11/13pt   *System* QuarkXPress™   [SE]

A catalogue record for this book is available from the British Library

Library of Congress Cataloguing in Publication data

Moore, J. (Janet)
Introduction to the invertebrates/Janet Moore; illustrations by Raith Overhill.
    p.   cm.
ISBN 0 521 77076 9 – ISBN 0 521 77914 6 (pb)
1. Invertebrates.   I. Title.
QL362.M66 2001
592–dc21   00-060818

ISBN 0 521 77076 9 hardback
ISBN 0 521 77914 6 paperback

Contents

List of Boxes

Preface

This book is an introductory guide to invertebrate evolution for university students and others. It is designed to be read as a whole, to interest and orientate students who encounter invertebrates. The book is needed because in our overcrowded biology courses sadly little is taught about invertebrates; again and again in recent years students have asked me 'Is there a little book?' because the current comprehensive invertebrate textbooks are too big and heavy for their requirements.

The book emphasises two questions about invertebrate evolution. Firstly, what invertebrates are there and how do they make a living? Secondly, what are the evolutionary relationships between the many and diverse animals that have evolved? *An Introduction to the Invertebrates* begins to give firm answers to questions about invertebrate relationships, in place of the speculation and controversies of the past. This advance is now possible, due partly to the finding of new fossils but mainly to information gathered very recently from new techniques of DNA analysis that have been used in the construction of molecular phylogenies.

The first two chapters of the book explain the process of evolution by natural selection and indicate how the pattern of evolution may be investigated by molecular as well as morphological methods. Each phylum is then considered in turn, since each consists of animals with a common body plan that will offer particular opportunities for evolution and impose particular constraints. After the body plan of each phylum has been defined, the ways of life open to animals so constructed are considered and the resulting diversity is indicated: this book therefore complements books discussing the animals living in a particular environment. Each chapter has a theme, indicated at the beginning of the chapter, relevant to the particular phylum. Boxes within the chapters present general background information about animal physiology, thereby avoiding interruption or repetition. The last two chapters, like the first two, are more general, discussing first animal development and then the evolutionary relationships revealed by fossils and by molecules. The text is illustrated throughout with newly prepared figures. Many of these are drawings of generalised examples and therefore no scales are indicated on the figures. The sizes of some particular species are mentioned in the text or the captions.

This book aims to discern and communicate overall patterns because the human mind cannot comprehend a scatter of unrelated facts. Classification is kept to a minimum, with just enough subdivision of phyla to provide group names needed for discussion; in general the aim is to use no more terminology than is necessary for communication. Generalisations are made that, on further enquiry, will require qualification, but this does not obviate the need for a framework of generalisations.

The book arises from many years of teaching undergraduates from many Colleges (including my own, New Hall) at Cambridge University. It is written in the hope that others will catch and share an enthusiasm for invertebrate animals.

Acknowledgements

It is a pleasure to thank those who have read and improved this book: firstly Dr Norman Moore and Dr Peggy Varley, who have read it all, and Dr Barbara Dainton, Dr William Foster and Dr Martin Wells, who have each read and helped with a number of chapters. For reading particular chapters I thank Professor Michael Akam, Professor Ray Gibson, Dr Liz Hide, Dr Hugh Jones, Dr Vicky McMillan, Dr Pamela Roe, Dr Max Telford, Professor Pat Willmer and unknown referees. I am grateful to them all for corrections and helpful suggestions – of course, remaining infelicities and mistakes are my sole responsibility. I should also like to thank Professor Ray Gibson for 30 years' happy collaboration in research on terrestrial and freshwater nemertines and Professor Pat Willmer for educating me even more than most of my past students did, and I extend that thanks to them all. I am grateful to Dame Rosemary Murray and New Hall for many delightful teaching opportunities, and to Professors Gabriel Horn, Malcolm Burrows and Michael Akam for hospitality in the Cambridge Zoology Department.

It is a pleasure also to thank Mr Raith Overhill for all the figures: sources for these drawings are acknowledged separately. At Cambridge University Press I should like to thank Dr Tracey Sanderson, for her help, guidance and encouragement throughout the writing of this book, Mrs Sandi Irvine and Mr Ward Cooper. Finally I particularly wish to record my gratitude to the late Professor Carl Pantin, my research supervisor, and Dr Barbara Dainton, my undergraduate supervisor, for teaching me and arousing my interest in Zoology, and to my husband Dr Norman Moore for sustaining and sharing that interest.

Illustration acknowledgements

All the illustrations were newly prepared by Raith Overhill, to whom I am most grateful. They were based either on original material or on illustrations in previously published works. While none of these has been reproduced directly, every effort was made to gain permission to reproduce illustrations from copyright holders, who are gratefully acknowledged as follows.

Blackwell Science: Barnes, R. S. K., Calow, P. and Olive, P. J. W. (1988) *The Invertebrates: A New Synthesis*: Figs. 3.59; 4.9; 4.14; 4.24; 4.25; 4.26; 4.29; 4.35; 4.41; 4.42; 4.45; 4.64; 6.9; 7.1; 7.17; 7.20; 8.12; 8.31; 8.32; 8.40; 8.41; 8.48; 8.49; and 16.20. Marion, M.A.F. (1886) *Archives de Zoologie expérimentale et générale* (2) **4**, 304–326: Fig. 7.6. Danielsson, D. (1892) *Norwegian North Atlantic Expedition* (1876–1878) *Rep. Zool.* **21**, 1–28; Carpenter, W.B. (1866) *Philosophical Transactions of the Royal Society of London* **156**, 671–756: Fig. 7.15. Also Barnes, R. S. K. (ed.) (1998) *The Diversity of Living Organisms*: Figs. 5.017 and 5.122.

Buchsbaum, R.: *Animals without Backbones* (1938) published by University of Chicago Press: figures on pages 79, 301 and 310.

Cambridge University Press: Borrodaile, L. A., Eastham, L. E. S., Potts, F. A. and Saunders, J. T. (1958) *The Invertebrata*: Fig. 408. Denton, E. J. and Gilpin-Brown, J. B. (1966) 'On the buoyancy of the pearly Nautilus', *Journal of the Marine Biological Association* **46**, 723–759: Figs.1 and 4. Trueman, E. R. (1975) *The Locomotion of Soft-bodied Animals* (Edward Arnold): Figs. 2.17 and 3.11. Young, D. (1989) *Nerve Cells and Animal Behaviour*: Figs. 2.2, 2.3 and 2.6b.

Company of Biologists Ltd: Gray, J. and Lissmann, H. W. (1938) *Journal of*

Experimental Biology **15**, 506–517: Fig.1. Gray, J. and Lissmann, H. W. (1964) *Journal of Experimental Biology* **41**, 135–154: Fig. 1. Weis-Fogh, T. (1973) *Journal of Experimental Biology* **59**, 169–230: Fig. 21A.

Elsevier Science: Based on Figures 13.7, 13.8 and 13.9 on pages 402–3, in Wolpert, L. *et al.*, *Principles of Development* published in 1998 by Current Biology Ltd, with permission from Elsevier Science.

English University Press: Chapman, R. F. (1969) *The Insects: Structure and Function*: Figs. 96; 286; 289 and 360A.

Garland Publishing: Alberts, B. *et al.* (1983) *Molecular Biology of the Cell*: Figs. 14.9 and 15.3.

Gibson, Ray: (1972) *Nemerteans* (published by Hutchinson): Figs. 6, 8A,B, 13H.

Harcourt: Barnes, R. D. (1974) *Invertebrate Zoology* (3rd edition) published by W. B. Saunders Co.: Fig. 3.4. Ruppert, E. E. and Barnes, R. D. (1994) 6th edition published by W. B. Saunders Co.: Figs. 3.10 (parts), 4.7, 4.8C, 4.15, 4.25, 4.31, 4.35,4.41, 4.46, 4.67, 6.4, 6.26C, 6.48, 8.47, 10.3B, 10.4A, 10.5, 10.6A, 10.12A, 10.19, 10.25, 10.57A, 10.58D,10.69,10.70,.10.73B, 10.82A, 10.88C, 11.6C, 11.13A, 11.22A, 11.23A, 11.50B, 11.63D, 11.66A, 12.2D,E, 13.4A, 13.8, 13.36A, 13.39A, 13.49, 14.6B, 14.8A, 14.11A, 14.30E, 14.78A, 14.85B, 15.1A, 15.6A, 15.11, 16.5D, 17.4B, 17.11C, 17.16A, 17.17A,17.19A, 17.35B,C, 18.9B, 18.19B, 18.43, 18.53,18.60A,19.5B, 19.10B, 19.27B.

Kluwer Academic Publishers: Wigglesworth, V. B. (1965) *Principles of Insect Physiology* (6th edition) Chapter 9 'Respiration' Fig. 228 (p. 322). Wells, M. J. (1978) *Octopus: Physiology and Behaviour of an Advanced Invertebrate*, Chapter 2 'Anatomy': Fig. 2.1 (p. 13). With kind permission from Kluwer Academic Publishers.

Reed Educational and Professional Publishers Ltd: Freeman, W. H. and Bracegirdle, B. (1971) *An Atlas of Invertebrate Structure*, reprinted by permission of Heinemann Educational Publishers, a division of Reed Educational and Professional Publishers: Fig. 50.

Weidenfeld and Nicolson: Wells, M.J. (1968) *Lower Animals* (published in the World University Library series): Figs. 4.7; parts of 4.8; 4.9; 5.5; 6.9; 9.6 B; 10.2 and parts of 10.4.

Worth Publishers: from *Animal Physiology* by Eckhart, R. and Randall, D. (1978) (W. H. Freeman & Company): Figure 9.8, used with permission.

1

The process of evolution: natural selection

This book is about invertebrate evolution. Every account of structure and function and the adaptation of an animal to its environment is a description of the results of evolution. Not only the intricate design but also the vast diversity of animals has been achieved by descent with modification due to the action of natural selection. A process so fundamental needs to be introduced at the very beginning of the book. As the different phyla are presented, general discussion of some other topics will become necessary (and will be inserted as 'Boxes') but evolution cannot wait.

1.1 What was Darwin's theory of natural selection?

Our understanding of evolution dates from the publication in 1859 of Charles Darwin's great book *The Origin of Species by Means of Natural Selection, or the Preservation of Favoured Races in the Struggle for Life*. Before that time, explanation of all the details of animal design in terms of a divine Creator was widely accepted, though perhaps the extraordinary variety of life (e.g. what has been termed 'the Almighty's inordinate fondness for beetles') was harder to explain. From very early times a few writers had postulated evolutionary theories, suggesting that different species might not all have been separately created, and further that complicated forms of life could have arisen from simple antecedents by descent with modification. This, however, was mere speculation in the absence of support from a large array of ordered facts. What Darwin gave us was a mass of careful observations, many gathered while he was Naturalist on the voyage of HMS *Beagle*, from which he formulated a theoretical framework showing that evolution could have occurred by what he called 'Natural Selection'. That time was ripe for such a theory is shown by the simultaneous conclusions of Alfred Russel Wallace from his work in Indonesia. The cooperation of Darwin and Wallace without any competition for priority is an encouraging example of decency transcending competition.

Darwin's argument was as follows:

1. Living things tend to multiply. There are more offspring than parents and, if unchecked, their numbers would increase in geometrical ratio.
2. The progeny cannot all survive, because resources (food, space, etc.) are insufficient.

Therefore there will be competition for survival, a 'struggle for existence' between individuals of the same species.

3. Living things vary; the progeny are not all identical and some will be better equipped for survival than others.

Therefore 'favourable variations would tend to be preserved, and unfavourable ones to be destroyed. The result of this would be the formation of new species' (N. Barlow, *The Autobiography of Charles Darwin*, 1958, p. 120).

To describe this process of natural selection Herbert Spencer used the phrase 'survival of the fittest'. The phrase needs to be qualified if misunderstanding is to be avoided: firstly, it is not mere survival but differential reproduction that is required and, secondly, 'fittest' does not refer to general health and strength but to some precise advantage in particular circumstances in a particular environment. Adaptation consists in the perpetuation of such an advantage down the generations.

Here at once was Darwin's greatest difficulty. For natural selection to work, advantageous changes had to be inherited. In Darwin's time heredity was assumed to involve the blending of the features of the two parents, and Darwin was much worried by the criticism (from an engineer, Fleeming Jenkin) that any system of blending inheritance would remove the advantage in a few generations. The solution was at hand, but never known to Darwin. Gregor Mendel had already shown that heredity was particulate, but his work was not publicised until 1900.

1.2 What was Mendel's theory of heredity?

Mendel's 'atomic theory' of heredity was based on his experiments on crossbreeding garden peas. He deduced that hereditary factors are constant units,

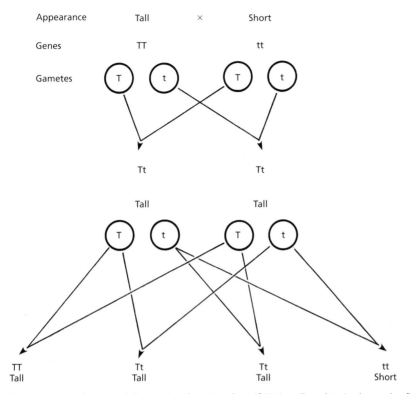

Figure 1.1. The Mendelian ratio: showing that, if 'T' is tall and 't' is short, the first generation will all appear tall and that (after self-pollinating) the second generation will have the ratio of 3 tall to 1 short.

handed down unchanged from parent to offspring, and that these units occur as 'allelomorphic pairs', the two members of each pair representing two contrasting characters. At sexual reproduction when gametes (spermatozoa and ova) are formed, only one factor of each pair can enter a single gamete. When gametes fuse to form a 'zygote' the factors, one from each parent, are combined. One factor in a pair may be 'dominant' over the other, which is called the 'recessive' and has no apparent effect on the organism but is maintained when it reproduces (Figure 1.1). The organism contains a very large number of such pairs (some 50 000 pairs in humans) most of which segregate and recombine independently at every sexual reproduction. Mendel's analysis explained both the basic resemblance between parents and offspring and the introduction of variation between them.

1.3 What is the cellular basis of heredity?

Early in the twentieth century, T. H. Morgan's studies of cell structure identified Mendel's factors as 'genes' borne on the elongated bodies, 'chromosomes', contained in the nucleus of almost every cell in the body (see Chapter 15, where the contributions of studies of the fruitfly *Drosophila melanogaster* are discussed). All organisms develop from the division of cells which previously formed part of one or (where reproduction is sexual) two parent organisms. A. Weismann first recognised that the 'germ-plasm' that gives rise to gametes is distinct from the rest of the body, the 'soma'. Somatic cells divide by 'mitosis', the longitudinal splitting of each chromosome with self-replication of each gene so that each half chromosome has exactly the same genes as its parent (Figure 1.2*a*): all the somatic cells in an individual are genetically identical. Gamete-forming cells first multiply by the same process of mitosis but then divide by 'meiosis', a process in which the number of chromosomes is halved and (usually) the two genes in each allelomorphic pair are separated as Mendel postulated (Figure 1.2*b*). The fusion of gametes combines half the genes of each parent to make a new individual. Gametes are described as 'haploid', since they contain half the number of chromosomes of the 'diploid' zygote and adult individual.

Only the gametes carry genes to the next generation. Changes often occur in the somatic cells, caused by use or disuse or by direct effects of the environment, but such changes cannot be transmitted to the offspring. Jean-Baptiste Lamarck is rather unfairly remembered mainly for his erroneous belief in the inheritance of acquired characters. Lamarckism has been typified by the idea that if giraffes stretched their necks to reach more food their offspring would be born with longer necks. A change to the body such as an elongated neck cannot directly affect subsequent generations: they can be changed only by the selection of individuals with genes promoting the growth of long necks. The 'phenotype', i.e. the organism defined by the characters made manifest, must be distinguished from the 'genotype' or genetic constitution, which alone can transmit changes to the offspring.

Note that the word 'develop' was originally used to describe two different consequences of gene action: the sequence of changes in an individual as the egg gives rise to the adult form, called 'ontogeny' and (on an enormously greater time scale) the process of evolutionary change called 'phylogeny'. We now reserve the term 'development' for ontogeny.

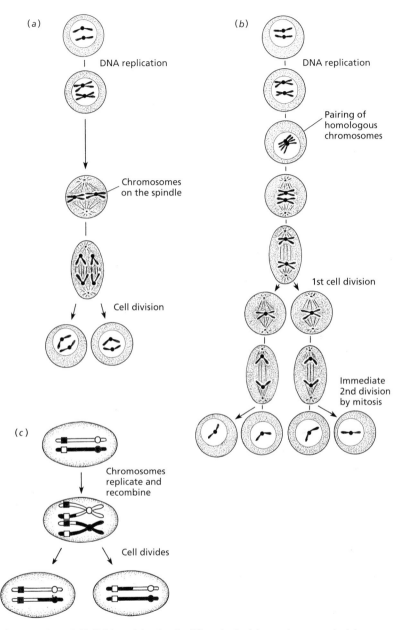

Figure 1.2. Cell division: (a) mitosis, (b) meiosis, (c) crossing over. In (c) squares represent genes, circles represent points of junction.

1.4 What is the origin of genetic variation?

Genes provide both the continuity and the differences between parents and offspring. The differences ('variations') are caused as follows:

1. Combination of half the genes from each parent.
2. Reassortment of the genes inherited from each parent. Genes borne along the same chromosome tend to be inherited together (they are said to be 'linked') but during meiosis there is normally some 'crossing over', or exchange of pieces of the split chromosomes (Figure 1.2 *c*).
3. The presence of a gene does not guarantee the appearance of the character with which it is associated, because gene effects may depend on the action of other genes present. The simplest example is dominance within an allelomorphic pair, but other genes may promote, suppress or alter the effect of a gene. A character may be the product of many different genes acting together, and one gene may affect many characters; for example, genes acting early in development may transform the effects of other genes acting later. It is a dangerous oversimplification to equate a character with the gene that in part governs it. Mendel was very lucky that his garden peas gave a simple picture.
4. Mutations occur. These may be chromosome changes, or more frequently errors in gene copying as cells divide. Sudden change in a phenotype due to mutation is rarely advantageous, as large changes tend to be lethal, but small changes may accumulate in the genotype, undetected until some change in circumstances gives them a selective advantage.

Clearly, mutation is the only one of these causes of variation that operates in asexual reproduction, where otherwise parent and offspring are genetically identical.

With the mechanism producing heritable variations understood, the picture of evolution caused by natural selection acting on random variations became firmly established. Due mainly to R. A. Fisher the emphasis fell not on the sudden change in form of an individual but on the spread of that variation through a population. The study of natural selection at work became a matter of statistics rather than qualitative descriptions. The synthesis of Mendelian genetics and natural selection was called Neo-Darwinism or 'The Evolutionary Synthesis', and by the 1930s it was widely accepted. It became the unifying principle underlying all branches of biology.

1.5 What is the nature of genes?

The work of James Watson and Francis Crick revealed in 1953 that DNA, in the form of a double helix, is the genetic material in the chromosomes. It replicates when the cell nucleus divides, and it can be transcribed to make RNA, giving a message that in turn can be translated into assembly of amino acids to make proteins. Genes and their action can now be studied at the molecular level, which has led to an enormous increase in understanding and opportunities for manipulation. There are also new problems: for example, some at least of the changes at the molecular level may not be due to natural selection. Could there be evolutionary change due simply to chance?

1.6 What is the role of chance in evolution ?

This has frequently been misunderstood by critics, some of whom regard the whole process as a combination of lucky accidents. They fail to distinguish the two stages involved. Variations arise by chance, as mutations and gene recombination occur at random. What is not at all a matter of chance is the operation of natural selection, which acts on these random variations to produce adaptation. Certainly change in gene frequencies may partly be due to chance; for example, long ago Sewall Wright pointed out that a small isolated population would contain only a few of that species' genes and therefore these genes would become overrepresented in the population. This process is called 'genetic drift' and cannot itself cause adaptation. It is still not clear whether 'neutral' (i.e. unselected) molecular evolution is important.

1.7 At what level does natural selection act?

Our understanding of evolution continues to evolve. Natural selection was at one time assumed to act for the good of the species or the group, until both experiment and theory showed that natural selection acts on individuals and cannot be shown to act on any larger entity. This at once produced new problems: a prominent puzzle is to find the advantage of sexual reproduction. Clearly this is slow and complicated compared with simple asexual multiplication, but of enormous benefit to the species because it introduces so much variation. How can this advantage apply at the individual level, and can an organism be unlike its parents? The problem is unresolved, but explanations focus on the masking of harmful mutations, the so-called 'Red Queen' effect:

the need to run as fast as possible simply to keep up. Host and parasite, for example, engage in a continual 'arms race': host offspring differing from their parents have more chance of avoiding their parasites, and parasites chemically different from their parents may evade the host's defences.

1.7.1 The unit of selection

Natural selection acts on the individual, but the effect of this action is the passing on of one set of genes rather than another. It is the relative frequency of genes that changes down the generations. As has been cogently argued by Richard Dawkins (in *The Selfish Gene*), the individual's body is the vehicle for the genes, which are the replicators; individual bodies are the genes' way of preserving the genes unaltered. Arguments about whether the individual or the gene is the 'unit of selection' are unprofitable: the important thing is to remember the role of each. Natural selection acts on the phenotype, not directly on the genotype. Further, no gene has a fixed selective value: its effect will depend upon other genes present. The danger of equating genes and characters must not be forgotten.

The above brief outline may serve to introduce invertebrate evolution, but further reading (see the end of text) is strongly recommended, to supply evidence for the above assertions and fuller discussion of these and many more facts and ideas.

1.8 What in general does evolution produce?

1.8.1 Diversity

This is the product of evolution. The very long evolutionary history of invertebrates has allowed an abundance of diversity: the very name 'invertebrates' is revealing: they can only be united as 'animals other than vertebrates'. Selection pressure caused divergence among the earliest multicellular forms, and certain body plans (called 'phyla', singular 'phylum') became successfully established, each offering particular opportunities and constraints for evolution, as this book will show. Within each phylum there are usually well-defined classes with characteristics that fit the animals for some particular environment or way of life, and within each class the original body form will have become modified as they exploit different habitats. This process of diversification from a single ancestral form is known as 'adaptive radiation'.

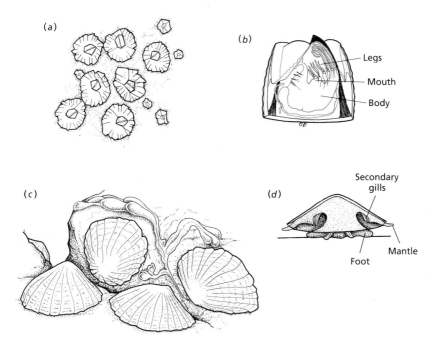

Figure 1.3. Barnacles and limpets on the seashore: (*a*) acorn barnacles; (*b*) vertical section through a barnacle; (*c*) common limpets; (*d*) vertical section through a limpet.

At the same time, natural selection does not only produce novelty: it also maintains and stabilises a successful structure and way of life. Nor does it only produce divergence: animals of very different ancestry may become very similar through adaptation when they solve the same problems or live in the same environment. Convergence is a widespread, often undervalued cause of resemblances, often baffling the attempt to classify animals by tracing their evolutionary history. For example, on an exposed shore the barnacles and limpets are superficially similar, being attached to rocks and covered by shells that protect them from desiccation and predatory birds (Figure 1.3). When the tide is up, the barnacle is seen to be a crustacean, shrimp like but attached by its head and kicking its food into its mouth with its legs, while the limpet is a mollusc with a muscular foot like a snail, moving off in search of food. This is a crude example: one has only to think of the mimicry in butterflies from different families to realise the fine degree of convergence that can be produced by natural selection, superimposed upon its primary divergent effect.

Natural selection defies our imposed categories: for example, in defining Platyhelminthes we state that the mouth is the only opening to the gut, yet one parasitic species has not just one anus but two. Animals are opportunists. Our categories especially meet trouble when we try to define a species, because we are trying to put firm boundaries on an evolutionary continuum. If species were incapable of changing, evolution could not have occurred.

1.8.2 Complexity

As well as diversity, complexity is a product of evolution. Primitively, multicellular forms were not very complex (note that 'primitive' means 'most like the ancestral form', not 'simplest'). What we call the 'higher' animals, those more recently produced, are on the whole very much more elaborate than their ancestors. The evolutionary pattern is clear, but must at once be modified by what we know of the evolutionary process. Complexity is not an end in itself: it will evolve where it has selective value, but not otherwise. Many simple forms survive today: one has only to look at sponges, animals extremely successful in that they are very numerous and widely distributed in the sea, yet remarkably simple in structure. A sponge is the best way of being a sponge, and natural selection has not in millions of years produced much alteration in their form. Cnidaria (anemones, corals, jellyfish, hydroids) again are simple in structure but remarkably numerous in the sea (unlike sponges they were also able to evolve great morphological diversity, as will be shown). In the more elaborate phyla the simplest animals are not necessarily the most primitive, as is clearly illustrated within the Platyhelminthes. Simplicity may be a secondary product of adaptation, and we cannot assume that a simple animal is primitive.

1.8.3 Not progress

Evolution is not directed from the outset and there is no inner directing force. The criterion for survival is immediate selective advantage, not any long-term evolutionary aim. We are being anthropocentric when we misapply the idea of progress to the evolutionary process. We like to think of all evolution leading up to humans at the apex of the evolutionary tree. This is a false picture.

1.8.4 Efficiency

This term needs careful definition. While natural selection should tend to maximise efficiency, that does not always mean maximum physical efficiency:

biological efficiency can be different. For example, a constant high body temperature enables the body's enzymes to work at maximum speed, which may be physically most efficient, yet it may be more advantageous to an animal to let the body temperature fluctuate, allowing the economy of cold inactive periods. Natural selection does not necessarily generate our own idea of a perfect product. 'Success', another anthropocentric approach, cannot be defined in terms of complexity or position in an evolutionary tree, but rather in terms of survival, abundance and perhaps also diversity.

As we study the invertebrate phyla that are the products of the evolutionary process, we can safely ask the question 'Why'? (as in 'Why is this animal so constructed?'). This is because we know that what a biologist means by such a question is 'How has such a structure conferred selective advantage'? Long ago when I was a student a professor said to us, 'When anyone asks me the question 'Why?' I refer him to a theologian.' I now think he was wrong – on both counts, because theologians cannot answer such questions and biologists can make the attempt.

2

The pattern of evolution: molecular evidence

2.1 How should we classify animals?

Classification is essential to any study of animals (the first attempt is attributed to Adam) and is a necessary prelude to tracing the pattern of evolution. Systematic ordering of the products of classification (taxonomy) can be done in various ways, but most usefully it aims to produce a 'natural' classification, i.e. a phylogeny that reveals evolutionary history. We try to put together those animals most closely related by descent, using resemblance as the basis for our classification.

Classification is difficult. A long-ago cartoon in the magazine *Punch* showed a railway porter scratching his head and saying 'Cats is dogs, and rabbits is dogs, but this 'ere tortoise is an Insect'. We must sympathise with his dilemma. The difficulty is that resemblance between animals is not an entirely reliable guide to their evolutionary history: it may be due either to close common ancestry or to convergence, occurring when animals of different ancestry acquire very similar adaptations because they face the same problems or live in the same environment. For example, any terrestrial invertebrate will have a skin relatively impermeable to water; this character is no guide to closeness of ancestry.

The main challenge in biological classification is to distinguish these two causes of resemblance, both of which are products of the evolutionary process. The great change in recent years is that we have new procedures of classification and new sources of information that may now enable us to trace invertebrate phylogeny. We need no longer agree with Libbie Hyman that when we attempt to relate phyla 'anything said on these questions lies in the realm of fantasy'. This chapter introduces our new tools.

2.2 How can we use morphology to trace phylogeny?

2.2.1 The traditional method

Animals are studied and compared. Characters indicating resemblance are picked out and assessed, as to whether they are independent of each other and whether the resemblance is likely to be due to convergence. Evidence is drawn from fossils, embryology, geographical distribution and any other available source. This process is subjective, yet it can be invaluable to use the opinion of an experienced systematist who has studied the particular group of animals. An advantage of this method is that careful assessment of characters is inescapable.

2.2.2 Phenetic taxonomy

'Phenetic' means 'as observed in the phenotype', and phenetic taxonomy originally meant 'classification by observed similarity' as in the traditional method. Phenetic taxonomy now frequently has a more restricted meaning, being equated with numerical taxonomy, a method where animals with the greatest number of common characters are put together. This process appears to be objective but there are concealed subjective steps: characters are selected and defined and then assumed to be finite and equivalent units, independent of each other. Assessment of the characters is vital, but the procedure does not ensure that it will be made. There is no attempt to identify convergence, which is assumed to be much less common than resemblance due to common ancestry and therefore likely to be eliminated when large numbers of characters are used. This method fails when convergence is common, or when only small numbers of independent characters are available.

2.2.3 Cladistic analysis

This is also called 'phylogenetic classification'; it groups organisms according to recency of common ancestry, as revealed by the presence of 'shared derived characters' i.e., when animals share the same difference from the primitive condition (Figure. 2.1*a*). For example, a primitive character shared by all insects, such as the possession of an exoskeleton, is not a character helpful for determining the relationships of a cockroach, a bee and a wasp, but the presence of hooks 'marrying' the two wings on each side of a bee and a wasp is a 'derived' (specialised) character, present in both, which does distinguish them from the

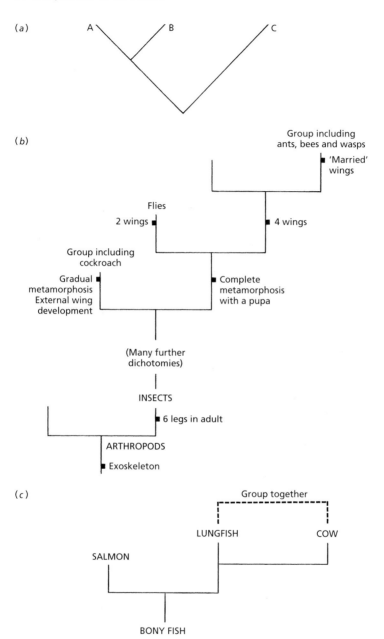

Figure 2.1. (a) Basic cladogram; (b) the cockroach, the bee and the wasp; (c) a hierarchy of recency of common ancestry.

cockroach (Figure 2.1*b*). The nodes (branching points) define differences from the primitive condition and these differences will be shared by all the lines beyond that node. These are the shared derived characters (sometimes called 'synapomorphies'). Cladistic analysis aims to identify 'monophyletic groups', i.e. the ancestor and all of its descendants. On the cladogram, such a group will be represented by a node and everything that follows from it. Time is not represented: the length of a branch has no significance. The sequence of common ancestry is all that is depicted (Figure 2.1*c*).

How reliable is such a cladogram? The emphasis on differences, which can be defined more precisely than similarities, and on the need to distinguish between primitive and derived characters, is very helpful. Much information is systematically obtained and made available for comparisons. There are, however, considerable drawbacks to cladistic classification, and now that the procedure is very widely adopted it is important to recognise these drawbacks. They are as follows:

1. Characters are too readily treated as fixed finite units that correspond in different groups, for example 'proboscis present' does not unite a worm and an elephant when 'proboscis' denotes totally different structures. This drawback can be overcome by sufficiently careful assessment of the characters used; the trouble is that cladograms can be drawn without such an assessment.

2. Rooting the cladogram (i.e. defining the primitive condition) is often very difficult. The root is compared with an 'outgroup', chosen as being related to the group but not part of it. Choice of the outgroup is subjective and difficult. It has reasonably been claimed that whenever outgroup analysis can be applied unambiguously it is not needed, and whenever it is needed it cannot be applied unambiguously.

3. The worst problem is that usually many possible cladograms can be drawn, and to pick the correct one the principle of 'parsimony' is invoked. Parsimony in this context means selection of the cladogram with the smallest number of evolutionary steps. The assumption here is the rash one that resemblance due to close common ancestry is much commoner than resemblance due to convergence, and the result is that perception of convergence is minimised. The difficulty of choosing between what may be a large number of equally 'parsimonious' cladograms is a further problem.

Despite these drawbacks, cladistic classification has been widely adopted and has revived interest in tracing animal relationships using morphological evidence.

2.3 How can we use molecules to trace phylogeny?

Recently, interest in phylogeny has received a new impetus, as our rapidly growing knowledge of genetics and molecular evolution is providing a new approach, a new source of evidence about the course of evolution. Molecules can be used in two different ways:

1. **Comparing the total genes of two different species**: the underlying idea is that if the genetic difference between two species is slight, they are liable to be closely related, and that the degree of genetic difference will indicate the closeness of that relationship. To base our study on the genotype itself seems very attractive, but we must as always be wary of reduction to the simplest units, as this will eliminate essential information that depends on the organisation and interaction of those units. The sheer presence of particular genes does not define their effects: the action of genes depends upon other genes present, and a small change in genes can make a big change in animals. That we share 98.4% of our genes with chimpanzees at once illustrates this point. Even though 1.6% may be a large number of genes, we are not as similar to chimpanzees as the figures suggest. The rate of genetic change is not the same as the rate of species change, and neither rate has remained constant during evolution.

2. **Using particular molecules in place of morphological characters**: here the idea is that if a molecule (a gene or the immediate product of a gene) changes slowly during evolution, comparison of the amount of change in that molecule will reveal the closeness of the relationship of different groups of animals. For example, a particular molecule might be compared in a range of insects, shrimps and spiders. Comparison should first show that different insects resembled each other more closely than any of them resembled shrimps or spiders, and might then indicate which two of these three groups are most closely related. In short, molecular characters can be substituted for morphological characters to assess the relatedness of animals, and it is this use of molecules that is discussed in the present chapter.

2.4 Which molecules are used?

Early work used the proteins, which are produced by gene action, but most of the recent work has concentrated on DNA, the genes themselves.

2.4.1 Ribosomal DNA

The genes coding for the RNA of ribosomes (rRNA), in particular for the small subunit (also called 18S rRNA; the gene is called *18SrDNA*) comprise ribosomal DNA (rDNA). This DNA was is very highly conserved, i.e. has changed very slowly in evolution, no doubt because it has an important structural role and mutations are unlikely to survive natural selection. It can therefore provide evidence about changes that occurred very early in animal evolution, such as the separation of classes within a phylum or even the origin of new phyla.

2.4.2 Mitochondrial DNA

The genes situated in the mitochondria, outside the nucleus and different from the nuclear genes, are useful sources of information at the other end of the evolutionary time scale. Mitochondrial DNA changes relatively fast in evolution and is used to resolve changes that occurred less than 15 million years ago, such as the separation of genera and species. In most animals the sperm contributes no material to the zygote, and therefore mitochondrial inheritance is confined to the female line.

2.4.3 Genes regulating early development

These may be very informative (see Chapter 19). Evidence obtained from the ordering of whole blocks of genes may avoid the disadvantages (see below) inherent in relying for phylogenetic evidence upon a single gene molecule, and therefore may provide a more accurate record of evolution.

2.5 How is molecular information obtained?

2.5.1 Gene products

The differences between proteins can be revealed and estimated using gel electrophoresis, and many other techniques were invented before the genetic material itself became accessible for direct study. Other techniques are now used more commonly.

2.5.2 DNA hybridisation

The paired strands of DNA dissociate when heated because the bonds between corresponding nucleotides are broken. They recombine on being cooled. Single strands of DNA from related species can be put together and bonds will form, but only at the sites which correspond. When such a 'hybrid' is carefully heated it dissociates at a temperature lower than that required to dissociate perfectly matched DNA, because fewer bonds will have been formed. The difference between the two temperatures ('melting points') can be used as a measure of the genetic similarity of the two species.

2.5.3 Restriction site analysis

Restriction enzymes cut DNA at predictable sites into fragments about 4 to 6 nucleotides long. Fragments from different sources can then be compared, to obtain information about a small part of the total molecule.

2.5.4 Sequencing of nucleotides

This process allows the identification of each nucleotide in the whole sequence of a DNA molecule. This exhaustive process has been made easier by modern techniques, especially the polymerase chain reaction (PCR), which amplifies a small quantity of material for rapid analysis, and by the automated sequencing machine.

2.6 How is molecular information processed?

Molecular evidence provides a large number of characters, all precisely defined. A character is usually a given nucleotide at a given site on the DNA molecule. Can the methods used for morphological characters be applied?

1. **The traditional method** of assessing characters is clearly not applicable.
2. **Phenetic analysis** can be applied to molecular differences. If molecular change increases at a constant rate as evolution proceeds, the amount of change is a measure of evolutionary distance. However, genes do not always change at a constant rate (see below).
3. **Cladistic analysis** is generally used, and some of its drawbacks disappear,

because the characters are precisely defined and equivalent, and sufficiently numerous for statistical analysis to be substituted for parsimony. Selection of an outgroup to root the cladogram is, however, even harder.

2.7 How reliable is molecular taxonomy?

This is a large and controversial question, but an indication of the advantages and disadvantages of using molecules can be given.

2.7.1 Advantages

1. The equivalence of data, since the nature and position of the unit is precisely defined.
2. The enormous size of the data set.
3. Statistical analysis of cladograms, avoiding the pitfalls of parsimony, is possible and only awaits agreement on the statistical methods to be used.
4. Where change in a molecule is rare, as in genes coding for ribosomal RNA, it becomes possible to trace relationships far back in time.
5. Non-heritable variation is avoided.

2.7.2 Disadvantages

1. The underlying assumption for most methods is that change in a gene will depend only on the mutation rate and the time elapsed, i.e. that an unvarying 'molecular clock' is ticking at a regular rate. However, the clock is known to be variable in certain conditions, and the whole idea of functionally neutral changes in genes is controversial. Some branches of the evolutionary tree are known to evolve very fast: should we compensate by a subjective decision to omit such species (or groups of species) from our calculations?
2. There is no record of past changes in characters. This is a serious disadvantage, as there are only four possible nucleotides for any site in the DNA molecule. If there have been changes from one nucleotide to another and back again, such 'multiple hits' cannot be detected.
3. There is no recognisable intermediate condition between characters and, worse, no primitive condition for a given site can be recognised.
4. Functional correlates of character change can very seldom be traced.

2.8 What is the present state of phylogenetic enquiry?

Ever since Darwin, biologists have wanted to understand the evolutionary relationships between groups of animals. Sources of evidence have included the fossil record and the study of animal development (ontogeny) as well as morphological comparisons. In recent years there has been rapid progress, due to new fossil discoveries, new understanding of the genetic basis of development, cladistic analysis of morphological characters and the use of molecular characters as an entirely new source of phylogenetic evidence.

Molecular characters are not better than morphological ones but they are different. They are copious and comparable with each other. Their primary value is in providing an independently derived phylogenetic tree for comparison with phylogenies based on morphology. Where possible, a number of different molecules may be used and the results compared. Molecular and morphological evidence sometimes suggest quite different patterns of evolution, but increasingly often the results coincide. When molecular data coincide with one morphological tree rather than another, this is strong evidence for the correctness of that tree.

It is time to describe the animals themselves, to explain the body plan of each phylum and the evolution that has occurred within it. Our present understanding of relationships between phyla is then summarised in the final chapter.

3

Porifera

Sponges are by far the simplest multicellular animals and are very different from all the others. They have no fixed body shape, no plane of symmetry and are covered in holes. All sponges live in water, nearly all in the sea. The cells are uncoordinated, cell differentiation is entirely reversible and cells may wander about in the background jelly. A whole sponge can be regenerated from a few separated cells. Sponges can almost be regarded not as individuals but as colonies of separate cells; almost but not quite, as most have a skeleton made of spicules that supports the body.

These very simple animals are none the less very successful and wide-spread: since the early Cambrian they have covered most of the suitable sur-faces on the shore and in the shallow sea. How is it that such simple animals can do so well? What has there been for natural selection to work on in this phylum? How fundamentally do they differ from other animals and what are their evolutionary and ecological relationships with them? To address these questions, we must study the basic structure and the different kinds of sponges and indicate the ways in which they make a living.

3.1 What are the distinguishing characters of sponges?

- **Sponges are sessile and immobile**, having neither nerves nor muscles. There may be slight contractility round the larger pores but it is very restricted.
- **Sponge cell types** are the distinctive collar cells or 'choanocytes' (Figure 3.1*a*), the 'pinacocytes' that make an outer layer and the 'amoebocytes'

wandering through the central jelly or 'mesohyl'. This jelly is needed for support; in contrast to other animals, neighbouring cells are not bound together by a basement membrane.

– **Small pores** perforate the whole body (the name of the phylum, Porifera, means 'pore bearing'). Water carrying food particles enters the body by many small pores ('ostia'), moved in by the beating of the flagella of the internal collar cells. These cells extract food particles from the water, which then flows out through larger pores, the 'oscula' (Figure 3.1*b*). The structure becomes elaborated during evolution, as in Figure 3.1*c,d*. This arrangement, with the principal openings exhalant, is unique to sponges.

– **The skeleton** is made of spicules (Figure 3.1*e*) of calcite (a calcium salt) or silica (a silicon salt) with or without a matrix of horny collagen-type protein. Such use of silica is a unique feature. An unusually wide range of skeletal materials occurs among closely related species of sponges.

3.2 What different kinds of sponge are known?

1. **Hexactinellida**, the 'glass sponges' (Figure 3.2*a*) are very different from other sponges. Most authorities believe that they evolved separately, in parallel with other sponges, and should be placed in a separate subphylum, if not phylum. They consist of a small group (400–500 species) in the deep sea (below 200 metres), with skeletons made from six-rayed spicules of silica.They are syncitial (the cells lack boundaries). There are no pinacocytes, no cells with any contractility and even the so-called choanocytes are not separate cells. Hexactinellids are orientated to the constant water currents of the deep sea.

2. **Calcarea**, with calcareous spicules. They occur in shallow waters (less than 100 metres). Examples are *Leucosolenia* and *Grantia* (Figure 3.2*b,d*).

3. **Demospongiae**, with siliceous spicules. Examples are *Halichondria* the breadcrumb sponge (Figure 3.2*c*) and *Spongia* the once commonly used bath sponge, in which there are no spicules but only protein fibres.

Subdivision of the Classes has traditionally depended on spicule structure, but more recently biochemical or reproductive characters, or those discovered by electron microscopy, have been preferred. For a time a further Class, the 'Sclerospongiae,' was proposed for some 15 species of sponges where spicules have become fused to make massive skeletons, building coral-like reefs since the Palaeozoic. This, however, is not a natural group; it includes some Calcarea and some Demospongiae that have evolved in parallel.

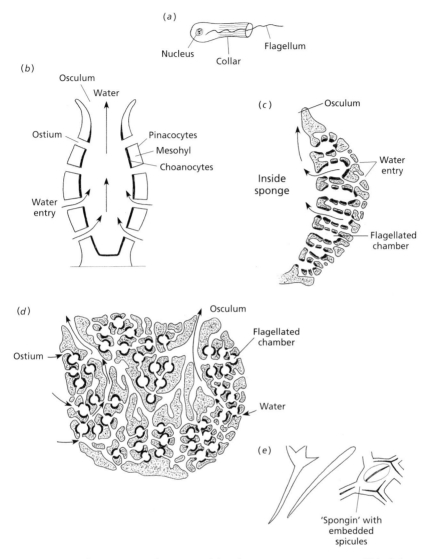

Figure 3.1. The structure of sponges: (a) a choanocyte, component of black layers in (b), (c), (d); (b) basic sponge structure, as in the late larva; (c) folding of walls to make flagellated chambers, for example *Leucosolenia* , one side; (d) a fully elaborated sponge, for example *Grantia*; (e) spicules.

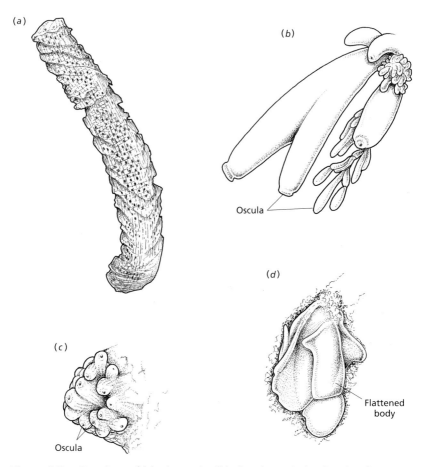

Figure 3.2. Drawings of (a) a hexactinellid, showing spicules fused to form a lattice; (b) *Sycon* (larger) and *Leucosolenia*; (c) *Halichondria*; (d) *Grantia*.

3.3 How do sponges make a living?

3.3.1 Feeding method

Choanocytes both create water currents and trap food. The collar is a ring of about 30 small folds ('microvilli', see Figure 3.1*a*), linked by cross-bridges. Food is absorbed by food-engulfing cells at the foot of the collar. Within both the Calcarea and the Demospongiae the internal pattern has changed (see Figure 3.1*c,d*) to provide canals and flagellated chambers. This arrangement provides a larger surface area covered in choanocytes and slows the flow of water past these cells, allowing more time for food capture.

Sufficient food particles arrive because water pumping is surprisingly intensive: for example, a specimen of *Leuconia* (Demospongiae) growing 10 cm tall with a diameter of 1 cm contained about 2.2 million flagellated chambers and pumped 22.5 litres of water per day. The oscular outward jet was 8.5 cm per second. Amoebocytes assist the circulation of food through the sponge: all digestion is intracellular. As in many permeable marine animals, dissolved organic material may be a subsidiary source of food.

3.3.2 Behaviour

The animal is sessile, there are no sense organs, nerves or muscles; what can there possibly be in the way of behaviour? Starting with the lowest expectations, we can find quite a bit. There is a measure of communication, even coordination, between cells. Stimuli such as touch, exposure to air or poisons can result in the closure of a distant osculum. Although there are no tissue junctions between cells, there may be communicating channels, rapidly and temporarily formed. Reactions are slow, as is shown by *Hymeniacidon*, a common encrusting orange growth, about a centimetre thick, on British beaches. Poke it, and about 10 minutes later the osculum will close. It is not clear how this contractility is achieved, but occurring round the oscula there are some amoebocytes called 'myocytes' that are particularly rich in microfilaments and microtubules. There is some evidence that myocytes may contain the fibrils (actin and myosin) that are the basis of contraction in all other animals investigated (and even in some unicells). Nor is it clear how cells can be re-extended, except by the pull of neighbouring cells, but water pumping must help to retain the shape of the sponge. After all, a sponge does not need rapid reactions: it needs only to close up fast enough to avoid desiccation when the tide goes out. A worse hazard would be to close the exhalant osculum while the flagella continued to beat: the sponge might burst.

It is in the very different Hexactinellida that a greater degree of coordination has been found. Here the diameter of oscula cannot change, but on mechanical or electrical stimulation the flagella in all the chambers may stop beating. Coordination has recently been shown to be electrical, although there are no nerves. In other sponges no such through conduction is known, but external currents can influence the rate of flagellar beat. In one experiment with *Halichondria*, in still water the flagella beat at 3 cm per second; this rose to 7 cm per second as the external current increased.

Animal skeletons, whether hard or hydrostatic (based on the incompressibility of water) usually translate muscle contraction into movement. Sponges,

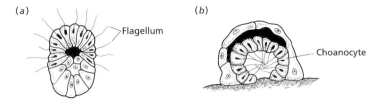

Figure 3.3. Sponge development: (*a*) amphiblastula larva; (*b*) gastrulation after settling.

however, have no muscles and the supporting skeleton instead serves to prevent movement from occurring. Spicules have further important functions in preserving the sponge's shape, keeping the pores open and maintaining the internal channels (as well as making the sponge even nastier to eat than it probably would be anyway).

3.3.3 Reproduction

Sponges have remarkable powers of regeneration: they can be strained through a fine mesh yet the cells will come together, aggregate and divide to reconstitute the sponge. Cells from different species will not aggregate. In the sponge body, asexual reproduction by budding occurs readily: it is hard to distinguish from growth. Some freshwater sponges bud off parts of the body to form 'gemmules', stages resistant to adverse environments that regenerate when conditions are favourable.

In sexual reproduction, gametes are formed in the mesohyl by dedifferentiation of other cells. Most sponges are hermaphrodites, but cross-fertilise. When sperm of the same species enters through an ostium, it is engulfed by a choanocyte which loses its flagellum and moves through the jelly until it finds an egg, a procedure very different from that of other Metazoa.

Sessile adult animals always need free-swimming larvae for dispersal. Sponges have simple flagellated larvae, usually developing in the parent body and then freed to swim and settle in suitable sites (Figure 3.3*a*). Some species achieve further dispersal by asexual fragmentation followed by release of larvae from the dispersed fragments. The settled larva gastrulates like other Metazoa (it used to be believed that sponge larvae turned themselves inside out, but that has been disproved by electron microscopy). The outer flagellated cells are lost at gastrulation and the inner cells move, divide and differentiate (Figure 3.3*b*).

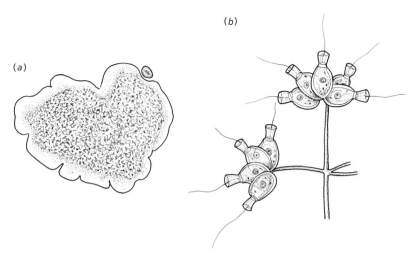

Figure 3.4.   Organisms apparently related to sponges: (*a*) *Trichoplax*, a placozoan (length 100 μm); (*b*) a choanoflagellate colony.

3.4 How are sponges related to other phyla?

This question has aroused much speculation. For a time sponges were separated from all other multicellular animals as a subkingdom, the Parazoa, and stigmatised as a 'side issue'. This classification and the ideas behind it have disappeared. Sponges are now known to be Metazoa, of the simplest possible kind, since the cells are so loosely held together. They are not very similar to any other phylum, with one small exception: Placozoa, flat plate-like aggregations of amoeboid cells, about 3 mm long (Figure 3.4*a*) formerly thought to be some developmental stage of a sponge. When sexually mature individuals were recorded, Placozoa were placed in a separate phylum (with one genus). They are probably not particularly closely related to sponges.

Sponge spicules fossilise readily. They are very common in Cambrian deposits and have recently been found among the very earliest fossil animals (see Chapter 20). Sponges cannot be classed with the Cnidaria as having two cell layers (nor are they radially symmetrical) and they are even less like all other Metazoa. Clearly no phylum can be derived from present-day forms, but choanocytes do resemble the unicellular Choanoflagellata, some of which form colonies (Figure 3.4*b*). This does indicate a possible line of metazoan evolution from flagellated unicellular organisms. Molecular evidence supports a single origin for the Metazoa, with the sponges as an early offshoot. If this is correct, choanoflagellates may be the sister group of all Metazoa.

3.5 What changes have evolved during sponge history?

3.5.1 Morphological change

In these simple animals with so little connection between cells, there is not much scope for morphological evolution beyond the elaboration of flagellated chambers related to increased efficiency of water circulation (see above). Sponges show an array of growth forms according to environmental conditions: where it is exposed to wave action a species usually displays a flat encrusting growth, but in crevices or still water the same species may grow tall or hang down, increasing the surface area. The continual disturbance of turbulent water brings in plenty of food, but in still water a sponge is in danger of recycling the water and gaining no more food. Increased body size is an advantage, and a raised osculum will achieve a more powerful exhalant jet. A few of the simpler Calcarea can be recognised by their shape; for example, Figure 3.2 shows the upright tubes of *Leucosolenia* and the flat 'purses' of *Grantia*, but usually the general appearance of a sponge is no certain guide to its identity. Deeper in the sea there is more variety, including metre-thick spheres and the reef builders already mentioned.

3.5.2 Physiological differences

Such differences may be marked and may be correlated with different ecological niches. In the Demospongiae for example, *Mycale* is an opportunist generalist readily colonising new sites but never growing to the maximum possible size; most of its energy is devoted to rapid growth and reproduction. *Tethya*, by contrast, forms permanent populations of large individuals in less favourable environments; its energies are channelled to physical resistance and it reproduces only slowly.

One extraordinary group of sponges has become carnivorous: living in deep sea habitats where small particle food is scarce, they capture small crustaceans with Velcro-like raised hooked spicules. The crustaceans become entangled, grown over and gradually digested. These sponges are hydroid-like in form (see Chapter 4) and have entirely lost choanocytes, ostia, oscula and water channels. They can be recognised as sponges (Demospongia) only by their spicules and by the nature of the outer layer.

However, the general form and function of animals so much constrained by structural simplicity did not give natural selection very much to work on. Diversity is shown far more strongly at the biochemical level.

3.5.3 Sponge biochemistry

Why are sponges so highly coloured? Not all their colours can be due to the commonly occurring symbiotic algae. On the shore and in the shallow sea they are often yellow, orange, red, green or violet. How can animals so openly exposed avoid predation, even if they do contain spicules? Why are they the only animals to be able to extract and build with silica? The answers to all these questions reside in unusual and very varied cell biochemistry. Colours may be due to pigment granules in amoebocytes and may serve as a warning of inedibility. Sponges produce an array of biotoxins that discourage predators; they may extend their use of poisons to chemical warfare with other sessile invertebrates, to compete for living space. One tropical encrusting sponge, *Terpios*, can grow as much as 23 mm per month as it poisons its neighbours. The family Clionidae (Demospongiae) includes 'boring' sponges with specialised amoebocytes whose chemical secretions remove calcareous fragments from coral skeletons, clams and scallops. The chips are collected into the exhalant currents within the sponge and pass out through the osculum. Boring sponges gain protective shelter – and cause considerable damage to coral reefs.

Cell biologists and pharmacologists are currently very interested in sponges. They provide model systems for the study of cell junctions of the simplest and most labile kind, and for the investigation of cell surface proteins that mediate cell recognition in a very basic immune system. Sponges produce bioactive compounds, some of which may directly benefit us: for example a sponge long used by New Zealand Maoris to promote wound healing has been found to contain high concentrations of a potent anti-inflammatory agent. A growing list of such examples emphasises that we need to conserve the biochemical diversity of sponges

3.6 How have sponges become so successful?

Morphological simplicity and lack of coordination have not prevented sponges from being extremely successful animals, if success is measured by survival, large numbers and very widespread distribution (in the sea). Sponges remind us that complexity of form is not the only route to success. Sponge diversity may be limited in morphological terms but their relatively independent cells have been able to evolve a variety of unusual biochemical specialisations.

Sponges are 'alternative' animals: they can respond to the environment and behave as functional units, but they do it in ways unique among multicellular animals.

4

Cnidaria

Cnidaria include the anemones, corals, jellyfish and hydroids, i.e. all the animals formerly included with the comb jellies (Ctenophora) in the phylum Coelenterata. They may be in the form of sessile polyps, or freely floating medusae (Figure 4.1). All are aquatic, nearly all are marine, and they are very simple in structure. Yet there are vast numbers of individuals belonging to at least 10000 species widely dispersed in the sea, varying in size from individuals a few millimetres across to coral colonies measuring hundreds of metres. Stinging cells (called 'cnidae' or 'nematocysts') are used for food capture and defence, and are unique to the phylum and diagnostic of it.

The combination of simplicity of structure with large numbers and considerable diversity provides the theme for this introduction to the phylum. Discussion is focused on three topics: how such simple animals can make a living, what features have enabled them to become so diverse, and how they might be related to other animals.

4.1 Why do we regard Cnidaria as simple?

They have no head end. The mouth (which serves also as the anus) is the single opening of the only internal cavity, called the 'coelenteron', which is an enclosed part of the water in which the animal lives. The mouth is usually surrounded by tentacles where the stinging cells are concentrated. Radial symmetry allows food capture from all sides, but it may be secondarily modified in relation to particular functional needs.

There are only two cell layers, the ectoderm (also called the epidermis) and

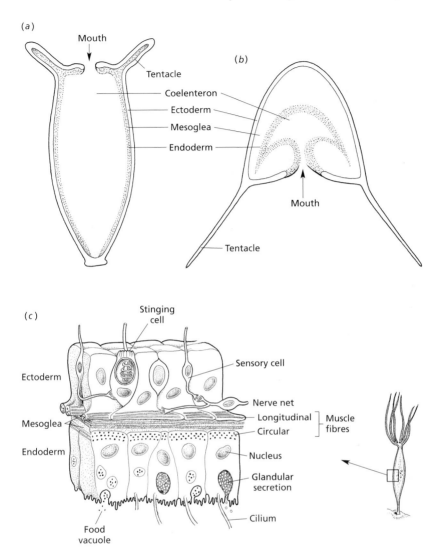

Figure 4.1. (a) Polyp; (b) medusa forms; (c) longitudinal section of *Hydra* to show cell types scattered in the two layers.

the endoderm (also called the gastrodermis). They are separated by a jelly-like 'mesoglea', which contains some cells and connective tissue fibres but is not itself a cell layer (Figure 4.2a,b). Accordingly Cnidaria are said to be 'diploblastic' in contrast to all other multicellular animals (except sponges), which are 'triploblastic', having three cell layers. Figure 4.1b shows that cells of the same type are not arranged together in either layer (the slightly confusing description

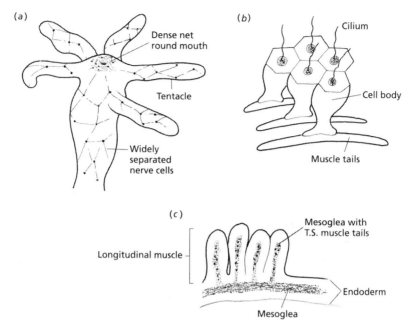

Figure 4.2. (a) The nerve net of *Hydra*; (b) muscle tails on the musculo-epithelial cells of a sea anemone, *Metridium*; (c) transverse section (T.S.) of mesentery showing folding of mesoglea with muscle tails, making a longitudinal 'muscle'.

of Cnidaria as having 'tissue grade' organisation emphasises that there is no aggregation of tissues to make organs). There is no brain or central nervous system, but a network of multipolar nerve cells conducts slowly in all directions (Figure 4.2*a*). There are no separate muscles but 'musculo-epithelial' cells of the ectoderm and endoderm are drawn out at the cell-base into contractile muscle tails that extend up and down or around the animal in the mesoglea. These muscle tails form sheets that may then be folded to make compact structures. This is a unique method of making solid muscles (Figure 4.2*b,c*).

As all cells in both layers are directly in contact with environmental water, either on the outside of the animal or in the coelenteron, there are no special structures for respiration and excretion, nor is there a transport system (apart from sea water channels in some large jellyfish).

Cell movement is a further distinctive property of Cnidaria. Not only are there migratory stem cells (called interstitial cells) that give rise to the nematocysts, nerve cells and gonads, but also apparently more differentiated cells may continuously undergo cell division and move. This capacity is most marked in hydrozoan polyps; in *Hydra*, for example, ectodermal cells are continuously produced just below the mouth region and migrate to the tentacle tips or to the foot, where they are sloughed off. At a certain distance from the dominant tentacular region, cells can move out and divide to form a bud. This ability to move and differentiate underlies the remarkable regenerative powers of many polyps; an isolated piece of tissue can often regenerate a polyp with one or more tentacular regions, even after having been put through a sieve, in strong contrast to what happens with more elaborate animals.

The simplicity of cnidarians is thoroughly established by their diploblastic constitution, the paucity of cell types, the nature of the only internal cavity, the lack of organs and the absence of centralised nervous systems and separate muscle cells.

4.2 How do Cnidaria make a living?

4.2.1 Nematocysts and feeding

Nematocysts (Figure 4.3) are the stinging cells that make it possible for these sessile polyps and floating jellyfish to be predatory carnivores, often feeding on animals larger than themselves. Nematocysts are specialised cells borne mainly on the tentacles. Each consists of a closed capsule covered by a flap and containing a coiled barbed thread or tubule filled with paralysing toxin. Some nematocysts do not contain toxin, the thread entangles small prey or adheres to it by the barbs. All this occurs within a fraction of a second. The used thread is discarded and interstitial cells move into the space to form more nematocysts. When capturing a brine shrimp *Hydra* may lose a quarter of its nematocysts, but replaces them within 48 hours.

Nematocysts are triggered by a combination of chemical and mechanical stimuli, causing the capsule flaps to open with a sudden release of bound calcium ions, which leave the capsule. Water then rushes in by osmosis. The greatly increased hydrostatic pressure inside the capsule, aided by release of stored energy in its walls, results in explosive discharge; the tubule is everted and toxin released from its tip into the prey.

Nematocysts can react quite independently, and as the whole mechanism is contained in a single cell they can react very fast. However, they are often coordinated by the nerve network, producing bursts of nematocyst discharge.

Nematocysts can also be used for defence against predators, and to attack rival polyp colonies competing for space. In some species of anemones and corals, contact with a foreign clone stimulates nematocyst discharge and the resulting 'stinging war' spaces out the different clones.

Nematocysts are not known in any other metazoan animals except for

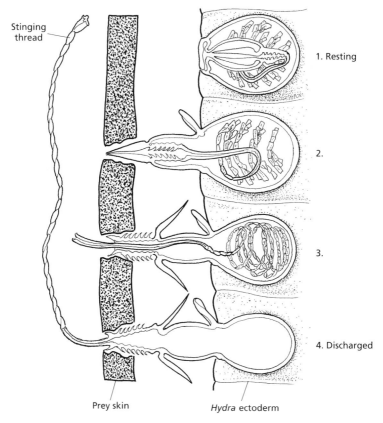

Figure 4.3. Nematocyst discharge.

some sea slugs, which extract them from their cnidarian food and, remarkably, harness them for their own use. A clue to the possible evolutionary origins of nematocysts comes from the spores of intracellular parasitic protista. Microsporidia have spores containing coiled eversible threads and Myxosporidia have spores containing nematocyst-like capsules with hollow eversible tubules. When in contact with a suitable host, the spore everts its tubule and through it discharges the entire spore contents into the host cell. Perhaps an early cnidarian ancestor was able to incorporate and use some such parasite, but that is pure speculation.

4.2.2 Nerve conduction

Response times in Cnidaria are slow, constrained by the structure of the nerve net and delay at the nerve cell junctions. Nervous centralisation is totally lacking. Where there are many nerve cells close together, as in the mouth region of a polyp, this is a region of particularly slow conduction rather than a rudimentary brain, as there are more junctions between cells that delay conduction (see Figure 4.2*a*). Apart from using the nematocysts, sessile polyps cannot make much response to stimuli and have no elaborate sense organs. Their nerve endings may be sensitive to chemical or mechanical stimulation, or to light, but it may be as much as 2 minutes before a cnidarian can react to a stimulus. Faster reaction occurs in some anemones (such as *Metridium*, often carried about on the shells of hermit crabs) where the column (Figure 4.4) has 'through conduction tracts' in the form of bipolar nerve cells up to 7 or 8 mm long. Here the first impulse 'facilitates' conduction so that subsequent impulses are not delayed at all. Again, some medusae have a nerve net in the 'umbrella' rim that stimulates muscle to pulsate rhythmically.

4.2.3 Movement and locomotion

Cnidarians can move parts of their bodies (mainly the tentacles) and change the body shape. Locomotion (movement from place to place) occurs in medusae and also in a few polyps. Movement by muscle contraction requires a skeleton, because muscles must have some restraint to pull against in order to have any effect, and after contracting they must be re-extended, but the skeleton need not be hard. Like other soft – bodied animals, Cnidaria use the incompressibility of water to serve as a hydrostatic skeleton. Hard structures exist in Cnidaria, such as the calcareous skeletons of corals, but they never play any part in muscle contraction.

Anemones (see Figure 4.4) and jellyfish (Figure 4.5) provide examples of cnidarian movement. A few anemones may shuffle, burrow or swim, but most are sedentary. They contract their muscles against the water-filled coelenteron, expelling water through the mouth: at rest the invaginated sleeve or 'pharynx' acts as a valve that keeps water from escaping. Refilling the coelenteron is slow; it depends on ciliated grooves or 'siphonoglyphs' running down from one or two points on the pharynx circumference. (This, incidentally, is an example of functional needs interrupting radial symmetry.)

In contrast to the slow reactions of anemones, jellyfish can swim relatively fast, as their muscles work against the much expanded mesoglea, which is elastic and springy. This 'jelly' is largely water; in solution the heavy sulphate ions of sea water are largely replaced by lighter anions, so that the mesoglea also provides the buoyancy that enables the animal to float. The bell pulsates rapidly due to the contraction of the striated muscle situated around its rim, and a group of nerve cells controls this rhythmical pulsation.

(a)

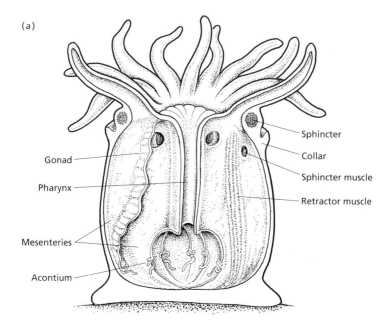

Sphincter

Collar

Gonad

Sphincter muscle

Pharynx

Retractor muscle

Mesenteries

Acontium

(b)

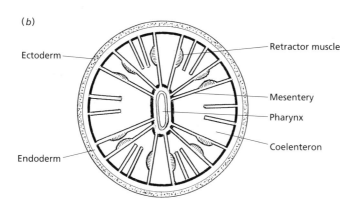

Ectoderm

Retractor muscle

Mesentery

Pharynx

Coelenteron

Endoderm

Figure 4.4.   The structure of a sea anemone (anthozoan): (a) longitudinal section; (b) transverse section.

(a)

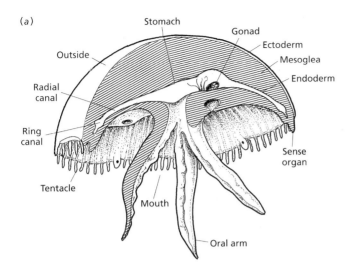

(b)

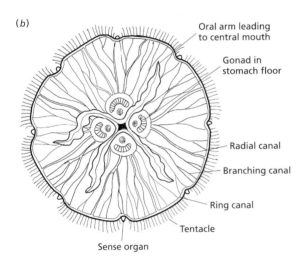

Figure 4.5. The structure of a jellyfish (scyphozoan): (*a*) side view, with shaded part in section; (*b*) oral view.

4.2.4 Spontaneous behaviour

How far this may occur in Cnidaria is partly a matter of definition. Certainly anemones frequently contract and alter their shape in the absence of any apparent change in conditions.

4.2.5 Reproduction

This may be asexual: as described earlier, *Hydra* polyps bud and the buds separate from their parent. The same process can result in the growth of a colony, if the buds do not separate. More commonly reproduction is sexual, achieved by the release of eggs and sperm into the sea.

4.2.6 Life history

Anthozoa have no medusa in the life cycle. The polyps release the gametes, which fuse to form a zygote and this develops into a ciliated swimming 'planula' larva, very simple in structure, which later settles and grows into the adult polyp. Some hydrozoans are similar (e.g. the freshwater *Hydra*) but characteristically the polyp colony, living in the harsh conditions of the intertidal region, buds off medusae, which produce gametes and the zygote develops into a planula, which grows into a new polyp colony (Figure 4.6*a*). This is not the 'alternation of generations' familiar in plants, since polyp and medusa have chromosomes identical in number and kind. Dispersal is of the utmost importance to a sessile animal, and the motile medusa provides a more robust agent of dispersal than the fragile planula larva. In Scyphozoa the gamete-producing medusa is dominant in the life cycle. The zygote develops not into a planula but into a sessile larva which buds off little jellyfish (Figure 4.6*b*).

4.2.7 Early development

Cnidaria, unlike other phyla, have a bewildering variety of developmental mechanisms. Such plasticity may be attributed to the basic simplicity of the structure and to the lack of constraints on cell movement. Differences in growth and development are found between closely related species and even between individuals within a species, depending on environmental conditions.

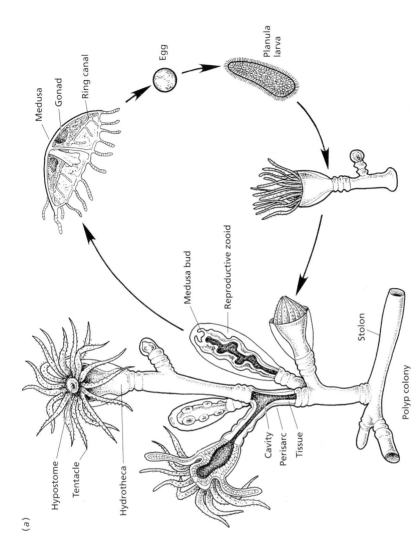

(a)

Medusa
Gonad
Ring canal

Egg

Planula
larva

Hypostome
Tentacle

Hydrotheca

Medusa bud

Reproductive zooid

Cavity
Perisarc
Tissue

Stolon

Polyp colony

Figure 4.6. Two cnidarian life cycles: (a) the hydrozoan *Obelia*, showing the structure of the polyp colony

(b)

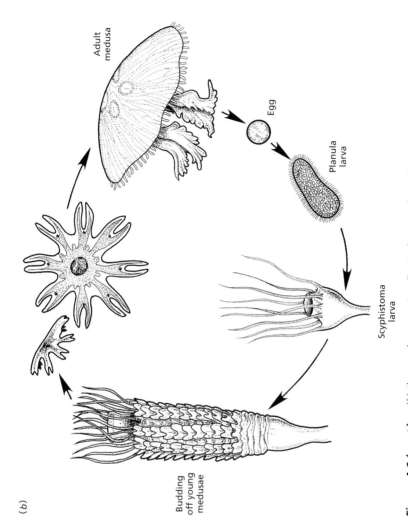

Adult
medusa

Egg

Planula
larva

Scyphistoma
larva

Budding
off young
medusae

Figure 4.6 (cont.). (b) the scyphozoan *Aurelia*. Both parts show the planula larva.

4.3 What kinds of Cnidaria are known?

There are four classes. The range of structure in cnidarians is shown in Figures 4.4 to 4.8.

— **Anthozoa**: anemones and most corals. Polyps with vertical divisions (mesenteries) in the coelenteron (see Figure 4.4). No medusa forms.

— **Alcyonaria**: polyps with eight mesenteries and typically eight branched tentacles; sea pens, branching corals, soft corals.
— **Zoantharia**: typically with six or 12 mesenteries and variable numbers of simple tentacles; anemones and 'true' oceanic reef-building corals.

— **Cubozoa**: box jellies. Medusae with four sides and a marginal shelf, or velum. In tropical seas, e.g. off Queensland, Australia, they are known as 'sea wasps' or 'stingers', and can be lethal to humans.
— **Scyphozoa**: jellyfish. Medusae with the mesoglea much expanded; no velum. The medusa form is dominant but may develop from a transient polyp-like sessile stage. Jellyfish are very common in all oceans; they may penetrate shallow seas or be washed ashore.
— **Hydrozoa**: hydroids, typically with both polyp and medusa stages in the life cycle. The polyps are small, without mesenteries in the coelenteron, and the medusa may have a velum. Polyp colonies are often intertidal; some groups are oceanic. *Hydra* lives in fresh water.
The many forms include:

— **Colonial polyps** (a few are secondarily solitary, e.g. *Hydra*).
— **Trachyline medusae** (oceanic, without polyps).
— **Milleporine corals** (see below).
— **Siphonophora** (highly specialised polymorphic colonies).

4.4 How has so much diversity been possible?

The diversity of Cnidaria is due to three attributes:

1. **Polyp and medusa forms**. These two forms provide the basic diversity of Cnidaria, since they make possible two different ways of life, a sessile polyp or a free-swimming medusa. As has been explained, many cnidarians combine both forms within one life cycle.
2. **Colony formation** is the commonest result of the remarkable budding power of many polyps, allowing increase in size and the possibility of

division of labour. Size increase is a great advantage in that a larger area can be swept for food. While individual medusae can enlarge by expanding the inanimate mesoglea (the largest jellyfish known are up to about two-thirds of a metre in diameter) individual polyps cannot get very big without supporting structures and increased digestive area; anthozoans with mesenteries dividing the coelenteron can become larger than hydrozoan polyps. In a colony, however, many small hydrozoan polyps can sweep as much water as a single anemone, and food is shared as the units remain in contact through a common coelenteron. They are limited in size by problems of mechanical support rather than by physiological constraints.

The commonest example of division of labour is that between the feeding and reproductive individuals in a hydrozoan colony. There may be further differentiation; for example, between feeding and stinging individuals in the colony. Such polymorphism reaches its height in the floating colonies of the Siphonophora, which consist of highly modified individuals where the distinction between polyp and medusa is obscured (Figure 4.7). The division of labour between individuals in these colonies resembles that between organ systems in more highly organised animals. Cnidarians do not have division of labour between organ systems in a body, but the siphonophores may have polymorphism between genetically identical individuals in a colony as a functional alternative.

3. **Coral formation** enormously extends the range and size possible for colonies. Any polyp that lays down calcium carbonate becomes a coral. Occasionally corals are solitary, for example the Devonshire Cup Coral, but nearly all are colonial. They occur in three groups of Cnidaria (Figure 4.8):

   — **Milleporine corals** (Hydrozoa): the skeleton is laid down outside the ectoderm of each polyp. The structure is never very large: it can be recognised by the two sizes of holes, large for feeding polyps and small for stinging ones. Medusae are produced at intervals (unlike anthozoan corals).
   — **Alcyonaria** (Anthozoa). The skeleton is laid down in the mesoglea by ectodermal cells. It may then be compacted, as in the Organ Pipe Coral, or there may be a central supporting rod made of protein or of coral itself, as in the red coral used in jewellery.
   — **Madreporaria** (Anthozoa). The skeleton is laid down underneath the polyps; living tissue from the polyp base extends into the skeletal mass, leaving its imprint between the calcareous septa. These are the 'true' reef-building corals.

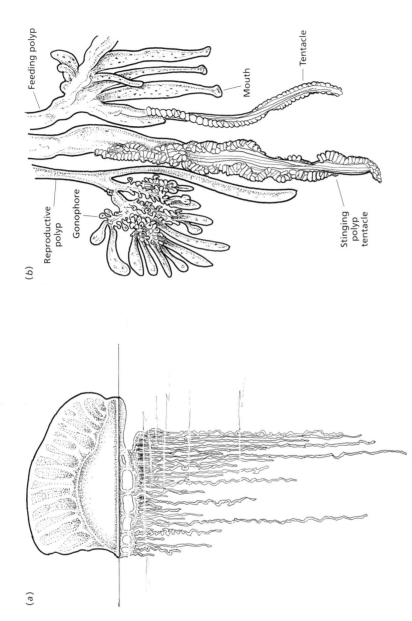

Figure 4.7. (a) *Physalia*, the Portuguese man-of-war, a siphonophore; (b) part of a *Physalia* colony showing the division of labour between individuals.

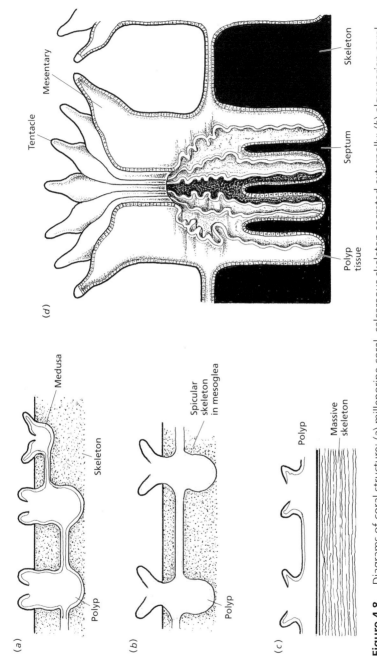

Figure 4.8. Diagrams of coral structure: (*a*) milleporine coral, calcareous skeleton secreted externally; (*b*) alcyonarian coral, skeleton laid down by ectodermal cells in the mesoglea; (*c*), (*d*) 'true' corals with massive external skeletons.

4.5 What is the ecological importance of coral reefs?

Reefs built by corals are the largest structures ever made by any animals, including ourselves. Deposition of large quantities of calcium carbonate by a colony of many small polyps depends on the presence of green algae, living inside the cells of the coral polyps; such an association for mutual benefit is called 'symbiosis'. Photosynthesis by the algae provides the coral with some of its food and the algae assist uptake of nitrates and phosphates, but their chief importance seems to be that photosynthesis facilitates calcification by removing carbon dioxide (CO_2). This promotes the dissociation of calcium bicarbonate dissolved in sea water, with formation and precipitation of calcium carbonate:

$$Ca(HCO_3)_2 = CaCO_3 + H_2O + CO_2$$

Coral reefs occur in tropical seas. They need sufficient light for photosynthesis by the symbiotic algae and a temperature range of 23–29 °C (with an optimum of 26–27 °C). These two factors also determine the depth at which living coral polyps can grow, This is often on top of the skeletons of their dead precursors because of past changes in sea level. Coral skeletons may extend for almost a kilometre and a half in depth, giving us a record of changes in sea level back to the Cretaceous era. Changes in sea level have enabled fringing reefs, barrier reefs and atolls (coral islands) to be formed.

Coral reefs are the basis of a whole ecosystem as they are inhabited by a great variety of fish, crustaceans, echinoderms and many other invertebrates. They have been called 'the rain forests of the sea' and support an even greater range of species (more than 93 000 have so far been described). Like rain forests they are greatly threatened.

Coral reefs are sadly being destroyed, partly by predators such as the Crown of Thorns starfish, but mainly by ourselves. Pollution, fishing by dynamite, rock mining and removing corals for sale all cause damage. Growing populations are building on fragile coral, and coastal deforestation causes blankets of sediment to wash on to the reefs and smother the polyps. Too frequently nowadays divers find that corals have been 'bleached', stripped of their algae, leaving only white skeletons devoid of life. Bleaching has many causes, but global warming resulting in rising sea temperatures is probably the main one. Major conservation efforts are urgently required to maintain this most valuable marine ecosystem.

4.6 How are Cnidaria related to each other and to other animals?

All four classes of Cnidaria can be traced back to Precambrian fossils but their relationships are controversial. Past candidates to be the most primitive group of Cnidaria have included hydrozoan medusae, hydrozoan polyps, anthozoan polyps and the transient polyp stages of scyphozoans. Modern methods do not help us greatly: morphological characters are few and variable, so that cladistic analyses give unconvincing results, nor does molecular evidence give us certainty. Mitochondrial DNA provides one important piece of evidence: in Anthozoa as in all other known Metazoa the mitochondrial DNA is circular (as in bacterial DNA) but analysis of 25 hydrozoans, 5 scyphozoans and 1 cubozoan revealed that they all have linear mitochondrial DNA, which is very likely to be a derived feature. This supports, but cannot prove, the morphological arguments suggesting that Anthozoa are nearest to the primitive cnidarians. Remaining uncertainty about the starting point does not help us to relate Cnidaria to other animals.

Molecular evidence supports the view that multicellularity arose only once in evolution, with Porifera, Cnidaria and triploblastic animals representing three separate multicellular lines. Attempts to derive triploblastic worms from particular groups of cnidarians have never been convincing, and suggestions that all cnidarians are degenerated platyhelminths are even less probable. Any animal could theoretically be derived from a planula larva, but that is merely a tribute to the planula's total lack of distinctive structure.

Cnidaria are therefore seen as separate from other animals, a monophyletic group that is an evolutionary dead end. These often very beautiful animals have unique features, and we can look at them with fresh eyes. They make their living differently from other animals, their basic simplicity of structure limits their behaviour but has allowed the evolution of great diversity, and they may dominate the marine environment, especially as corals.

5

On being a worm

Any soft-bodied legless animal whose length exceeds its width is liable to be described as a worm, and many invertebrates fit this description. Four of the main worm-like phyla are discussed separately in later chapters, but there are many other different worms, belonging to phyla often castigated as 'minor', usually because they have a small number of species or are very small animals. This chapter introduces the variety of worms, after considering why worms should have evolved so many times and what muscular machinery is necessary for their locomotion.

5.1 Why are there so many different kinds of worm?

Mechanical facts about the molecules that make up animals mean that worms are very easily produced. Cells secrete extracellular compounds with charged molecular backbones: like charges repel, causing linear extension, and linkage between these large molecules provides orientation in a structure that will be anisotropic (i.e. have different properties in different directions). If a blob of soft tissue has such orientated fibres, any event such as growth or motion or external pressure will automatically turn that blob into a cylinder. Orientated fibres will guide and limit the direction of growth, and enable it to change its shape. No further genetic instruction is needed to make a worm, in its simplest form.

A worm, then, is easily produced: why should such a structure be favoured by natural selection? Soft tissues are extraordinarily resistant, and damage is readily repaired. There is a large range of possible sizes, from less than a millimetre (as parasites or in the marine 'interstitial' habitat in between sand grains) to over 30 metres in the sea. Movement in one direction will be favoured by worm-like shape: an anterior end becomes established, usually with at least a simple ' brain' and sense organs, and the distinction between dorsal (top) and ventral (underside) surfaces confers bilateral symmetry. Very small worms can be propelled by cilia, but most worms move by muscle contraction (see Box 5.1). An important difference from Cnidaria is the presence of a third tissue layer, the mesoderm, separating muscle fibres from cells of the other two layers and allowing them to run in all directions.

5.2 How can muscles move a worm?

5.2.1 Hydrostatic skeletons

Water is incompressible: a closed bag of fluid can change its shape but not its volume. Soft-bodied animals lacking hard skeletons use this fact to re-extend their muscles. The earthworm is a familiar example: longitudinal and circular muscles contract alternately acting against the fluid in the body cavity, bristles grip the substrate and the worm moves forward by the process called 'peristalsis' (Figure 5.1*a*).

5.2.2 Body cavities

Some worms have no body cavity between the gut and the outer surface: they are said to be 'acoelomate', and all that the muscles have to work against is the fluid pressure of water held inside cells or in spaces in the soft tissues. A much more effective hydrostatic skeleton is provided by a fluid-containing body cavity. Box 5.2 introduces the different kinds of body cavity and Box 5.3 shows where they originate and how they develop.

Examples of the use of hydrostatic skeletons in invertebrates include:

– **Burrowing** (many worms). Circular contraction extends the anterior end, which takes a hold, then longitudinal contraction pulls up the posterior part .
– **Wave motion** (nematodes, polychaete annelids and many others). Waves of contraction are propagated alternately along the sides of the worm, producing an **S** shape, with backward pressure on the environment so that the worm moves forward (Figure 5.1*b,c*).

Box 5.1
Muscle

Muscles contract and relax: they cannot actively stretch. Therefore every animal must be able to re-extend its muscles before they can contract again. Much of animal design depends on this simple fact. Muscles often occur in antagonistic pairs (as in the muscles either side of our limb joints) where contraction of one

(a)

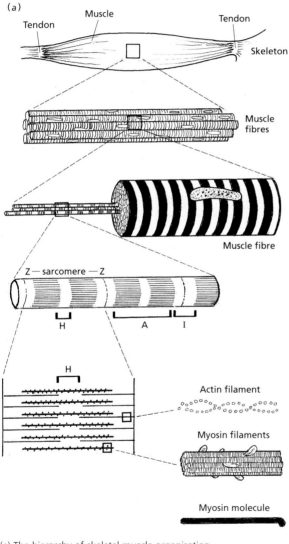

(a) The hierarchy of skeletal muscle organisation.

Box 5.1 *(continued)*

(*b*)

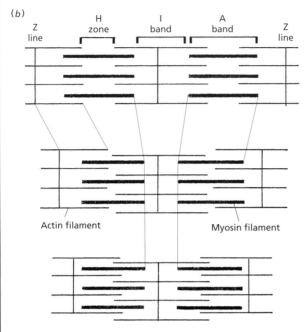

(*b*) Shortening of muscle by sliding filaments: the unit which shortens is the sarcomere, between two 'Z lines'. Myosin (thick) filaments are shown dark. The muscle appears striated because myosin-containing regions (the 'A' bands including the 'H' zones) alternate with regions containing actin filaments only (the 'I' bands).

extends the other. To have any effect, muscles must work against resistance in the form of a skeleton, which may be hard (as in ourselves and insects) or hydrostatic, based on the incompressibility of water (see Chapter 5). Invertebrates have evolved a great range of musculature within the constraints of these basic requirements.

Muscle fibres contract by shortening and by building up tension: some muscles ('isotonic') change mainly in length, while others ('isometric') change little in length but greatly in tension. All contraction depends on the sliding of microscopic filaments within the fibres (see diagram). Thin actin filaments slide between thick myosin filaments with formation of cross-bridges between them and at the same time energy is released to power the contraction. The tension developed is proportional to the number of cross-bridges. This universal machinery also makes muscle elastic, to a variable degree.

The following kinds of muscle can be distinguished in invertebrates:

> – **Striated muscle** appears under the microscope to be cross-striped because the filaments are closely packed, with their darker regions (where actin and myosin overlap, see diagram) coinciding. This fast contracting, elastic, often isometric muscle is found throughout the animal kingdom, from pulsating bells in jellyfish to vertebrate voluntary muscle.
> – **Insect asynchronous flight muscle** is a special case of striated muscle, being the fastest and most isometric muscle known. Contraction is faster than the nerve input, enabling some small insects to beat their wings over 1000 times per second (see Chapter 15).
> – **Helical smooth muscle** may appear plain or may show stripes in a spiral round the fibre if the bands coincide. This is the commonest invertebrate muscle. A variant is 'long-fibred' smooth muscle.
> – **Paramyosin muscle** has very long and large filaments, as in the adductors of clams. Slow to react, very isotonic and inelastic, it can exert very much more force and sustain it for longer than any other muscle.
> – **(Smooth muscle** is a vertebrate specialisation. Slow and very isotonic, it enables internal organs to maintain tension at very different degrees of extension.)

- **Jet propulsion** (some jellyfish; octopus, squid and cuttlefish).
- **Muscular waves** (platyhelminths, the foot of snails) effect slow propulsion.
- **Parts of animals** (spiders' legs, starfish tube feet) may work by muscles squeezing fluid, even though a hard skeleton exists.

5.2.3 Lattices

All hydrostatic skeletons need the support of connective tissue lattices: threads of collagen, incompressible in length, are wound helically round the soft body (Figure 5.1*d*). The animal can therefore contract evenly without bulging or kinking as the lattice angle changes. The lattice is a spirally coiled fibre like a spring, which on stretching becomes thinner, i.e. the lattice angle becomes smaller, and on compression becomes fatter, increasing the lattice angle. At one extreme the lattice angle tends to zero, at the other to 90°, but in an animal the stiffness of the bounding layer would prevent the angle becoming very small or very large.

Graphical representation of lattice action considers the worm as a cylinder. Then

volume = length × cross-sectional area (T.S.)

and the T.S. is proportional to the diameter. Since the fibre of which the lattice is made cannot change in length, the volume and the lattice angle will

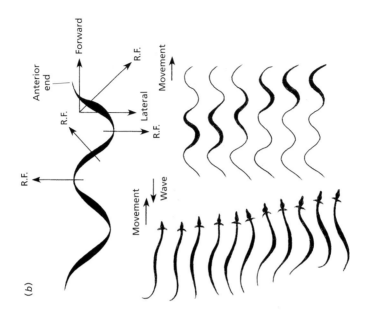

(a)

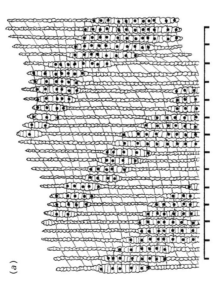

(b)

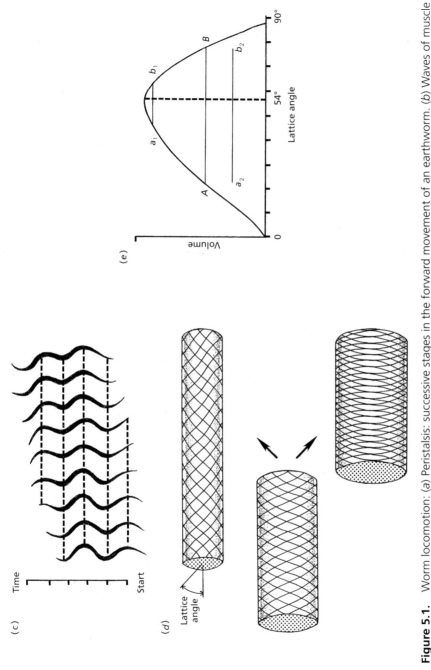

Figure 5.1. Worm locomotion: (a) Peristalsis: successive stages in the forward movement of an earthworm. (b) Waves of muscle contraction passing from the anterior to the posterior end of an undulating worm; forces generated (above) and resulting locomotion (below). R.F., resultant force. (c) The nematode *Haemonchus contortus* creeping over the surface of an agar gel. Tracings from successive photographs at 0.3 second intervals. (d) Connective tissue lattices, showing how the spirally arranged collagenous fibres control the effect of muscle contraction. (e) Graph to show the relationship between volume and lattice angle in a cylinder. *AB* represents most worms, a_1b_1 any terrestrial planarian or nemertine, a_2b_2 a marine nemertine, *Lineus longissimus*.

be related as shown in Figure 5.1*e*. Since a worm does not change in volume but in length, it must be represented on this graph by a horizontal line such as *AB*. *A* and *B* represent extremes of the lattice system: the worm cannot be longer and thinner than it is at *A*, nor shorter and fatter than at *B*. The line of the graph defines the only positions where the lattice allows the worm to be circular in cross-section: the worm must therefore flatten (becoming oval in T.S.) as it changes between *A* and *B*. Different worms will have different positions of *AB* on the graph; a land-living worm, for example, will tend to be circular in cross-sectional since this shape presents a minimal area for water evaporation, while some swimming worms can become flattened (Figure 5.1*e*).

5.2.4 The disadvantages of a hydrostatic skeleton

A hydrostatic skeleton has the great advantage of being resistant to impact damage: worms do not readily break, buckle or burst; distortions are not harmful; and considerable changes in shape are allowed. There are, however, formidable disadvantages. Firstly, the skeleton depends upon hydration. To be amply supplied with water the animal must at least be physiologically aquatic (i.e. confined to damp places if terrestrial). The weight of fluid required offsets the advantage of not having a heavy hard skeleton (the threads of the lattice are so strong that they can be very thin and light). Secondly, to move the whole body there must be a great deal of muscle, unlike animals with legs for leverage where muscle can be concentrated for efficiency. Further, in a crawling worm there will be a large surface impeded by friction on the substrate. Locomotion is even more uneconomical in energy requirements because the whole body must be accelerated and decelerated all the time. Thirdly, such a system is much harder for nerves to control, and impossible to control precisely: a beetle (for example) can know exactly where its legs are and the degree of bending at each joint, but no such information exists for a worm. Quite apart from the vulnerability to predators of a soft body, the advantages of being a worm should not be overstated. All the same, the range of worm phyla is impressive.

5.3 What worm phyla are known?

The four main phyla are described in Chapter 6 (Platyhelminthes), Chapter 7 (Nemertea), Chapter 8 (Nematoda) and Chapter 9 (Annelida). This chapter introduces the smaller phyla illustrated in Figures 5.2 to 5.4, with brief text indicating for each phylum the number of known species (very approximately), the size of the animals, their habitat, and distinguishing features.

Why include all these 'minor phyla' in an introduction to invertebrates? Partly to show the range of body plans that have evolved in 'worms', partly on account of their great relevance to our consideration of phylogenetic relationships. In the past, these phyla were grouped according to the presence or absence of a body cavity; for example, acoelomates with platyhelminths, coelomates with annelids. All the phyla diagnosed as 'pseudocoelomates' (see Box 5.2) were combined with nematodes as 'Aschelminthes'. Increasing knowledge about the probable evolutionary relationships of these animals has not only abolished the 'Aschelminthes' but also established very different groupings, and some of the 'phyla' named below may not be independent. To avoid confusion, the present chapter maintains the more traditional identity and arrangement of these phyla. Chapter 20 presents the modern evidence for a different phylogeny.

– **Mesozoa** (Figure 5.2*a*. 50 spp. Length about 0.5 mm, marine). Acoelomate endoparasites with complex life cycles and extremely simple structure: an outer ciliated layer, very few non-reproductive cells in the inner (mesoderm) layer and no endoderm. The phylum includes two different groups, the Rhombozoa (dicyemids) and the Orthonecta.

– **Gnathostomulida** (Figure 5.2*b*. 100 spp. Length <1 mm, marine). Acoelomate miniature anaerobic flatworms gliding through the sand, often very numerous in the interstitial habitat. Their name derives from the specialised muscular pharynx with jaws which grasp food.

– **Gastrotricha** (Figure 5.2*c*. 450 spp. Length <0.5 mm, marine). Pseudocoelomates. Like the Gnathostomulida they glide through often anoxic sand or mud. They have external cilia and a protective cuticle, not strengthened as in nematodes but with adhesive tubules. The cilia sweep bacteria and other food into the mouth.

– **Acanthocephala** (Figure 5.2*d*. About 1000 species. Length up to 800 mm, aquatic). Specialised endoparasites, the larvae living in insects and crustaceans and the adults in the gut of vertebrates. There is a proboscis covered in hooks, a pseudocoelom and no gut. The surface cuticle has conducting channels.

– **Rotifera** (Figure 5.2*e*. 2000 spp. Length up to 3 mm, mostly in fresh water). Pseudocoelomates. They are propelled by a crowning 'wheel' of cilia, the corona. The ciliated epidermis has an intracellular 'lorica' that can form a resistant resting stage in unfavourable conditions: this coupled with remarkably rapid reproduction in favourable conditions has

Box 5.2
Body cavities

A body cavity is a fluid-filled space lying between the gut and the outer body wall of an animal and containing the major organs. Possession of a body cavity allows an animal to increase in size, and may provide a hydrostatic skeleton against which muscles can contract. Cnidaria have no body cavity, the coelenteron is an enclosed part of the outside world, but it may serve as a hydrostatic skeleton.

Acoelomates

Those with no body cavity include:

Platyhelminthes	Mesozoa
Gnathostomulida	Nemertea

There are essentially three kinds of body cavity.

Haemocoels

The primary body cavity or blastocoel (see Box 5.3) persists to some extent in most protostomes, either enclosed within narrow blood vessels as in annelids or open as a blood-containing space called a haemocoel. Rapid blood circulation is then difficult to achieve, but all the tissues are continuously bathed in blood. Haemocoels occur in Mollusca and Arthropoda.

Pseudocoels

These are cavities with outer mesodermal layers but no lining cells between the cavity and the gut. A pseudocoel may be a persistent blastocoel (i.e. a haemocoel without blood) or possibly derived from vacuoles within cells, and It functions in much the same way as a coelom. It occurs in:

Nematoda	Rotifera	Kinorhyncha
Gastrotricha	Acanthocephala	Priapula (?)
Nematomorpha	Loricifera	

Coeloms

These are secondary body cavities bounded on all sides by mesoderm. They may be formed as schizocoels or enterocoels (see Box 5.3). They serve primarily as hydrostatic skeletons but they also allow the gut and body wall to move

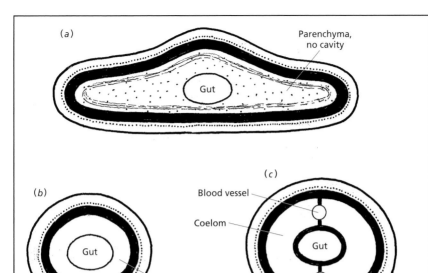

(a)

Parenchyma, no cavity

Gut

(c)

(b)

Blood vessel

Coelom

Gut

Gut

Pseudocoel
or Haemocoel

Transverse sections: (a) acoelomate, (b) pseudocoelomate, (c) coelomate. Muscle, derived from mesoderm, is shown black.

independently and may be sites of transport, excretion, and storage of gonadial products. They occur in:

Annelida	Pogonophora	Brachiopoda	Echinodermata
Sipuncula	Mollusca	Phorona	Chordata
Echiura	Arthropoda	Bryozoa	Chaetognatha

made the phylum very successful. Parthenogenesis is common; the class Bdelloidea never has any males and sexual reproduction is unknown.

– **Nematomorpha** (Figure 5.2*f*. 250 spp. 'Hair worms', may be 360 mm long and only 1 mm in diameter. Freshwater or damp soil, with one marine genus). Like nematodes they have a thick outer cuticle, a pseudocoelom and only longitudinal muscle, and move by propagated waves. The young are parasitic in arthropods and the adults do not feed.

– **Loricifera** (Figure 5.3*a*. 10 spp. Length <0.3 mm, marine). Discovered in 1983, these marine interstitial burrowers have an introvert (eversible head), a trunk (with a pseudocoelom) encased in a lorica and the larvae have 'feet' posteriorly.

– **Kinorhyncha** (Figure 5.3*b*. 150 spp. Length <1 mm, marine). These very

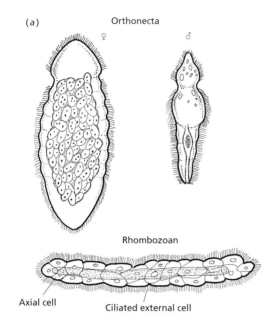

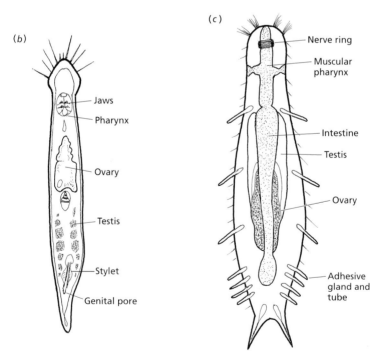

Figure 5.2. The diversity of worms: (a) Mesozoa from the two subgroups: orthonectans, female and male, and a rhombozoan (dicyemid); (b) a gnathostomulidan; (c) dorsal view of a gastrotrich, seen as if transparent;

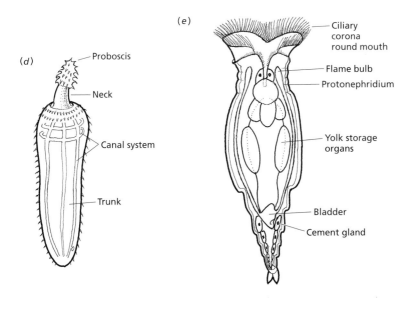

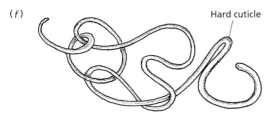

Figure 5.2 (cont.). (d) an acanthocephalan; (e) a bdelloid rotifer; (f) a nematomorph.

small carnivorous worms are pseudocoelomates remarkable in being thoroughly subdivided into 13 parts, the first of which is an introvert used for burrowing in mud. This subdivision resembles the 'segmentation' found in annelid worms (see Chapter 9) but kinorhynchs are not otherwise similar to any segmented animals .

– **Priapula** (Figure 5.3c. 16 living species, 11 fossil. Length up to 60 mm, marine). They are also carnivores burrowing in marine mud, using the conspicuous introvert. The body is covered in a chitinous cuticle. The body cavity may be a true coelom not (as previously thought) a pseudocoel. They have long been seen as an ancient, isolated phylum.

Pogonophora and Echiura are customarily presented as separate phyla, but both may have arisen together within the Annelida:

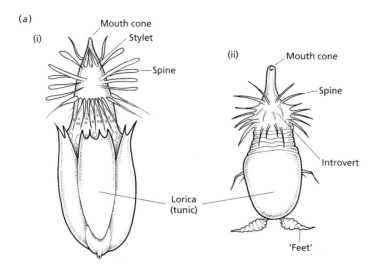

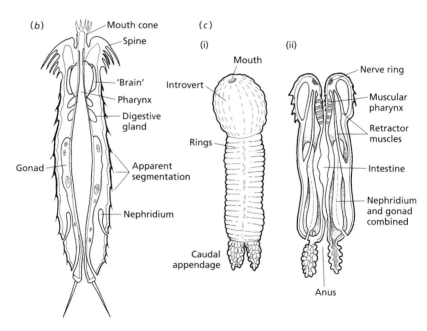

Figure 5.3. The diversity of worms (continued): (*a*) dorsal view of (i) adult and (ii) larval loriciferan; (*b*) a kinorhynch, showing the internal anatomy; (*c*) the (i) external morphology and (ii) internal anatomy of *Priapulus*.

– **Pogonophora** (Figure 5.4*a*, marine) are most unusual in that they have no alimentary canal (gut). They are segmented coelomate worms secreting tubes of chitin and protein in which they live, sticking upright from the ocean floor. Tissue replacing the gut is packed with sulphur bacteria that are the main source of food, but also the outer surface of the worm absorbs dissolved organic matter. The two subphyla are:

  – **Perviata** (95 spp. Length of tubes up to 8000 mm) occur in soft sediments to a depth of 9000 metres. They need reducing sediments, which supply reduced sulphur compounds to their bacteria, overlain by oxygenated water supplying oxygen to the worm's haemoglobin.
  – **Vestimentifera** (15 spp. but the growing exploration of ocean depths rapidly increases this number. Tubes may be 2 metres long and 30 mm in diameter). They are plentiful in the deep ocean vents where all life is based on sulphur bacteria. The trunk below the tentacles (gills) bears wing-like 'vestiments' in place of the bristles occurring in Perviata. Juvenile vestimentiferans have recently been found to have guts, lost in the adult.

– **Sipuncula** (Figure 5.4*b*. 320 spp. Length 1–300 mm, marine). They are sedentary deposit feeders, having tentacles, a retractable introvert and a trunk. Unsegmented coelomates.
– **Echiura** (Figure 5.4*d*. 150 spp. Length up to 500 mm, marine or brackish). Sedentary deposit feeders known as 'spoon worms', as sediment is collected by the elongated proboscis projecting from the trunk. Unsegmented coelomates.
– **Chaetognatha** (Figure 5.4*c*. 90 spp. 3–120 mm long, marine). Chaetognaths are carnivores, major predators in the marine plankton. These 'arrow worms' are streamlined, with rigid transparent bodies and horizontally projecting fins. They swim very fast and pounce on their prey, seizing it with the two groups of movable spines projecting on either side of the mouth. The worms are coelomate with longitudinal and oblique muscles contracting against the hydrostatic skeleton, a well-developed nervous system and haemal sinuses in place of a circulatory system. They are hermaphrodites with direct development.

   Chaetognaths are a small and uniform group, but often very numerous as individuals. In the English Channel two species are very common, *Sagitta elegans* from the Atlantic and *S. setosa*, which is an indicator species for coastal Channel water. At Plymouth in Devon an increased proportion of *S. elegans* indicates an influx of Atlantic water, overlying the Channel water, because this species has gas-filled pockets in the intestine causing the worm to float higher in the sea.

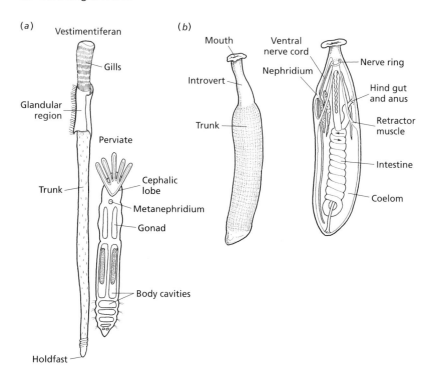

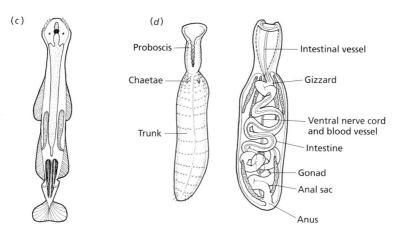

Figure 5.4. The diversity of worms (continued). Some coelomates:
(a) Pogonophora: a vestimentiferan and a perviate; (b) a sipunculan, external view
and internal anatomy as shown by dissection from the left side; (c) external view of
the chaetognath, *Sagitta elegans*; (d) an echiuran, external view and internal
anatomy shown by dissection from the dorsal side.

Box 5.3
Protostomes and deuterostomes

Chapter 19 gives an account of development, but an introduction is needed sooner, because early development may be a key to evolutionary affinities as well as to adult form (the ambiguous word 'development' is here applied to individual change, **ontogeny**, rather than to evolutionary change, **phylogeny**). Developmental patterns depend in part on the quantity of yolk in the egg: yolk is invaluable to feed the embryo but it is heavy and inert, impeding the division and movement of cells in which it is contained. The following account applies primarily to animals (such as most marine invertebrates) where the eggs contain little yolk and usually hatch early into feeding larvae.

Phyla may be divided into two large groups according to alternative methods of development at three early stages:

1. cleavage, the first divisions of the fertilised egg,
2. the fate of the blastopore, the earliest opening of the future gut,
3. the formation of mesoderm, and (in animals which have one) the coelom.

These two groups are:

– **Protostomes**: All the phyla with three cell layers (triploblasts) except
– **Deuterostomes**: echinoderms, chordates (and a few others? see Chapter 16).

Cleavage

A non-yolky fertilised egg typically divides into two cells and then each divides at right angles into two cells lying in the same plane. The third division may be either **radial** (Deuterostomes), with cells directly on top of each other or **spiral** (Protostomes), where cells are placed between those underlying them (see diagram). Cleavage continues, and in either case produces a ball of cells or **blastula**, hollow because the cells secrete fluid into the centre, forming the **blastocoel** or primary body cavity.

The Blastopore

Next, cell movements called gastrulation produce a two-layered structure, where an inner endoderm approaches the outer ectoderm, thereby much reducing the blastocoel. The endoderm-enclosed cavity is called the 'archenteron' or future gut, opening by a hole called (rather misleadingly) the blastopore. This hole becomes the mouth in protostomes and the anus in deuterostomes (so-called because the mouth is a secondary opening, elsewhere).

Mesoderm

In protostomes the future mesoderm is determined very early in cleavage and is identified in the early larva as solid blocks of cells either side of the posterior end of the gut. These cells then split or hollow out to form a 'schizocoelic' coelom.

Box 5.3 (continued)

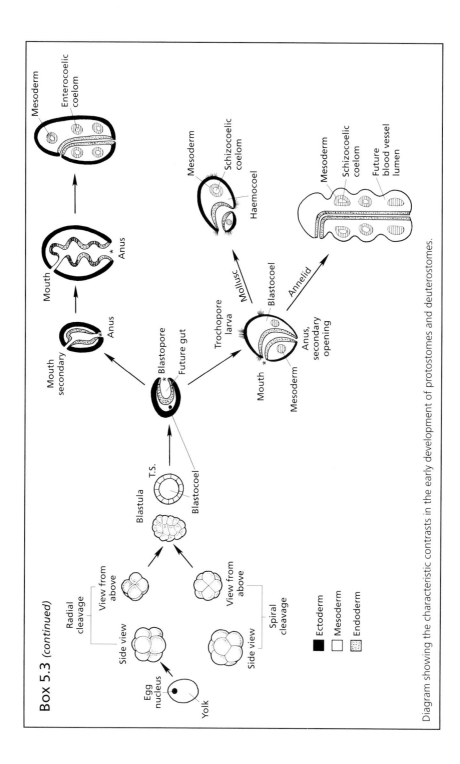

Diagram showing the characteristic contrasts in the early development of protostomes and deuterostomes.

The diagram opposite illustrates these processes both for an annelid (segmented; coelom enlarged and almost obliterating the blastocoel) and for a mollusc (unsegmented; coelom very small, leaving a large persistent blastocoel). In deuterostomes mesoderm is formed by out-pouching of sheets of cells in the gut wall; the pouches separate, each enclosing a part of the archenteron, which becomes an 'enterocoelic' coelom. In summary:

Developmental Stage	Protostome	Deuterostome
Cleavage	Spiral	Radial
Fate of blastopore	Mouth	Anus
Mesoderm and coelom	Early determined schizocoelic	Gut out-pouching enterocoelic

That three distinct characters in early development unite to distinguish protostomes and deuterostomes is a clear indication of evolutionary relationships. The division is generally accepted, and supported by molecular evidence. The categories are not, however, as clear and absolute as the above formulation suggests: there are examples of radial cleavage combined with a protostome mouth, and anomalous species can be found in most phyla. Again and again in Biology, there is a clear framework but exceptions abound, because natural selection is operating.

While chaetognaths possess some features of deuterostomes, such as radial cleavage and a coelom divided into three regions, other features are anomalous and chaetognaths are no longer placed among deuterostomes. The relationships of this isolated phylum remain uncertain.

5.4 Do Ctenophora belong among the worms?

On shape, symmetry and method of locomotion the answer is clearly that ctenophores are not worms, but this isolated small phylum has features that may further define a worm by contrast.

– **Ctenophora** (Figure 5.5*a*. 100 spp., marine). They are superficially medusa – like, bags of jelly floating and swimming in the sea. They are widespread and abundant (as individuals rather than species) in the oceans, with new deep sea forms currently being discovered. Closer inspection, however, shows that they are not very like the Cnidaria (traditionally their companion 'coelenterates') and they lack cnidae. Their main unique feature is the fusion of large cilia into 'comb plates' arranged

(a)

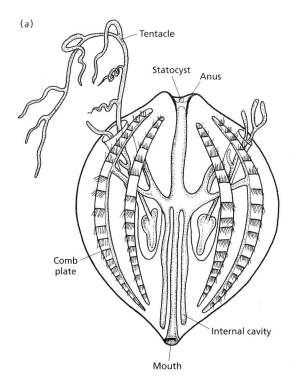

Tentacle

Statocyst

Anus

Comb plate

Internal cavity

Mouth

(b)

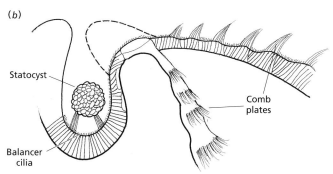

Statocyst

Comb plates

Balancer cilia

Figure 5.5. Ctenophora: (a) lateral view of a typical ctenophore; (b) aboral end of a ctenophore showing statocyst and comb plates;

transversely across each of the eight 'comb rows', coordinated by the apical sense organ from which they radiate (Figure 5.5*b*). This machinery enables ctenophores to swim more efficiently than small animals using normal cilia, although they lack the muscular development of a worm.

Ctenophores are not radially symmetrical, since all but one small group

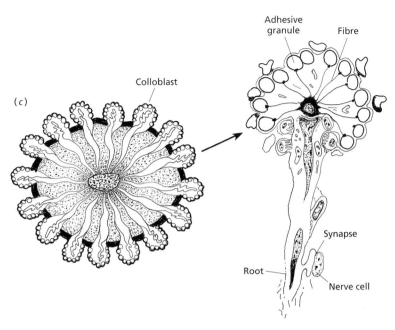

Figure 5.5 (cont.). (c) section through a tentacle of a ctenophore with detail of a colloblast.

bear a pair of tentacles and have digestive canals following this 'biradial' pattern: nor are they bilaterally symmetrical like a worm, since they lack an anteroposterior axis with a brain in front. Unlike cnidarians they have a third cell layer, containing a network of smooth muscle cells separate from the inner and outer layers. Ctenophores are able to be pelagic predators because they have unique adhesive cells, called 'colloblasts' or lasso cells (Figure 5.5c), but these are not at all like cnidarian stinging cells. The mouth is used for ejection of faeces, but there is a separate anus for removal of metabolic waste.

Ctenophores have unique cell types – ciliated comb cells, colloblasts, giant smooth muscle cells – and (as electron microscopy reveals) unique cell junctions. Their isolation is emphasised by their early development: the first cell divisions form a unique biradial plate and subsequently a distinctive larval form occurs. Altogether, ctenophores are not particularly close to cnidarians, nor indeed to any other phylum.

6

Platyhelminthes

Platyhelminths are present-day animals with ancient origins. The same can be said of any invertebrate phylum with living representatives, but the platyhelminths are particularly likely to be misunderstood. Reasonably called 'flatworms', they have been too readily equated with the ancient flatworms from which nearly all more elaborate multicellular animals were thought to have been derived. As soft-bodied animals leave at best an incomplete fossil record, for morphological evidence we turn to the simplest modern forms in our search for the earliest (i.e. most primitive) metazoa. Sponges and cnidarians are fundamentally unlike other animals: they have no head end and are not bilaterally symmetrical, nor do they have three cell layers (see Chapters 3 and 4). Other simple phyla are specialised parasites. Platyhelminths remain in the limelight as simple representatives of mainline multicellular animals. Yet they have a very long evolutionary history, and closer examination reveals distinctive specialisations that occur regularly (but never universally) in modern members of the phylum. How far is it misleading to consider platyhelminths as present-day representatives of ancestral flatworms?

To address this question, the following account first shows that the simplicity of the basic structure makes it difficult to trace relationships within the phylum and that there are many specialised features in the free-living platyhelminths as well as in the parasitic species. Relationships within the phylum are then discussed, as a basis for assessing its position.

6.1 What is the body plan of the phylum?

The basic structure of platyhelminths is very simple (Figure 6.1 *a–c*). They are 'triploblastic', i.e. have three cell layers, outer ectoderm, mesoderm and inner

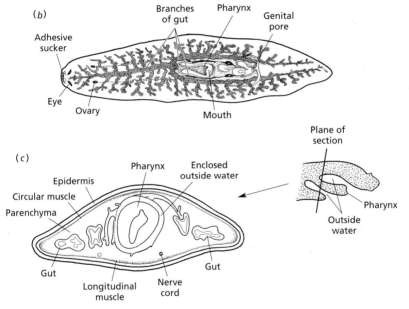

Figure 6.1. (a) The structure of a planarian (length about 1 cm, width 0.2 cm); (b) longitudinal section of planarian showing branching gut; (c) transverse section of planarian in pharyngeal region.

endoderm. In the free-living forms there is a head end with anterior sense organs and a rudimentary brain. They are commonly regarded as primitive on account of a number of negative characters: the mouth is the only opening to the gut, they have no body cavity, no respiratory system, no blood system, no appendages and no hard skeleton. Free-living platyhelminths move by means of the cilia or by contraction waves in muscle working against fluid pressure of the body contents (gut, parenchyma, genitalia, etc.). Animals without a respiratory or a blood system are necessarily flat, being limited by the diffusion of oxygen, and, since they are soft bodied with the length greater than the width, they are called worms.

6.2 What groups of worms constitute the Platyhelminthes ?

— Free-living platyhelminths with ciliated epidermal cells. They are the Turbellaria: they live in the sea, in fresh water or in damp places on land. This is a very heterogeneous group including Orders with a range of primitive characters (the Nemertodermatida and Catenulida) and the Rhabditophora, which is a large Superorder (Figure 6.2).
— Parasitic platyhelminths. Most platyhelminth species are parasitic (see BOX 6.1). There are three Classes:

  — Monogenea, flukes ectoparasitic on aquatic animals (mostly fish).
  — Trematoda including Digenea, flukes endoparasitic in vertebrates.
  — Cestoidea, tapeworms, highly modified endoparasites.

In all these forms, the epidermis is replaced by a 'tegument' protecting the parasite and governing what it can absorb from its host.

6.3 Is simplicity a reliable guide to phylogeny?

The interpretation of morphological simplicity is a major problem when one is seeking relationships within or between phyla. Simplicity of structure in a modern animal may indeed denote direct descent from the primitive condition, but it may alternatively be simplicity that has been secondarily evolved. The parasitic platyhelminths, as will be seen, include striking examples of secondary simplification, and simplicity is a very poor guide to relationships among the free-living members of the phylum. According to one theory, the platyhelminths have evolved by reduction from larger animals with a body cavity such as the coelom. As will be seen, the evidence is against this, but it does have to be considered. There is also a third possibility: a group of modern simple animals may be derived from an ancient simple animal not on the direct ancestral line. An outstanding example may be provided by the Acoela, a group that has been included among the simplest free-living platyhelminths but may have evolved quite separately.

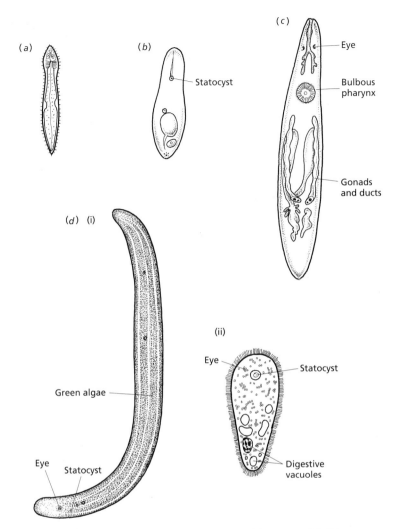

Figure 6.2.   Some free-living platyhelminths: (*a*) a catenulid; (*b*) a
nemertodermatid; (*c*) a rhabdophoran, all lengths about 1 cm; (*d*) an acoelan,
Convoluta roscoffensis (i) much extended to show symbiotic green algae, and
(ii) condensed and diagrammatic. Length about 0.3 cm.

6.4 Should the Acoela be placed in a separate phylum?

The Acoela are very small and simple (Figure 6.2*d*). They have no hollow gut,
but the innermost parenchyma is a digestive layer that may only form after
feeding. This parenchyma may be 'syncitial', i.e. with nuclei scattered in the

cytoplasm and lacking cell boundaries. They were once thought to be entirely syncitial, and were considered by some to be the earliest Metazoa, derived directly from the multinucleate but unicellular Ciliata, until electron microscopy revealed cell boundaries elsewhere in their bodies. Acoela show simplicity also in the structure of the nervous system; on the other hand they have a number of curious features not found in other Metazoa: there is hardly any intercellular substance, the structure of the spermatozoa, the cilia and their rootlets, and the protonephridia is very unusual. Accordingly, they have been described as secondarily simple.

However, analysis of the ribosomal RNA genes of Acoela does not support the view of Acoela as a specialised group. When some unusually rapidly evolving acoelans and others had been excluded, comparison of the 18SrDNA of 18 acoelan species and 74 other metazoan species showed that the Acoela separated out at the very base of the Metazoa, distinct from all the rest, suggesting that they are very primitive. Strong supporting evidence that they are different from platyhelminths comes from their early embryonic development: this had long been known to be unique (see Chapter 19), a fact that has acquired new significance. Possibly this chapter should be headed 'Platyhelminthes and Acoela': their relationships are under debate.

This obscure group of small worms can teach us a great deal about the search for evolutionary relationships. Simplicity may be a shared primitive condition or may be due to secondary simplification, but there is a third possibility: it may be a product of convergent evolution from a separate primitive condition. Molecular evidence may help us to distinguish between these possible causes of simplicity.

6.5 What is specialised about modern platyhelminths?

6.5.1 Reproduction

Ancestral flatworms probably resembled other simple marine invertebrates in having separate sexes, shedding sperm and eggs (without much yolk) straight into the sea and hatching early as small feeding larvae. Most modern platyhelminths by contrast have most elaborate systems of muscular glandular organs forming extremely complicated hermaphrodite reproductive systems (Figure 6.3a). Most turbellarians live in fresh water, and as is typical in this habitat they have internal fertilisation of large yolky eggs, which develop directly without delicate larval stages. Parasitic forms have even greater problems. They have

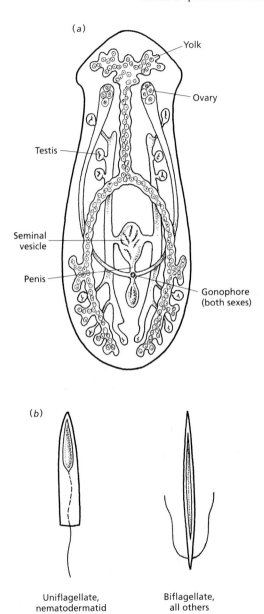

Figure 6.3. Reproduction in Turbellaria: (a) the hermaphrodite reproductive system; (b) turbellarian sperm.

even more elaborate reproductive systems and a series of larvae that facilitate transfer between hosts (see below).

Spermatozoa provide a striking example of specialisation in the Turbellaria. Most animals have the familiar 'tadpole-like' sperm moved by a single flagellum, but nearly all turbellarian spermatozoa have two flagella. The characteristic cap or 'acrosome' is absent, the nucleus is at the anterior end and the two flagella extend through the lateral cytoplasm of the cell (Figure 6.3*b*). There are exceptions: catenulid sperm have no flagella at all, and nemertodermatids alone among turbellarians have external fertilisation, and here the spermatozoa are uniflagellate, though highly modified from the primitive condition. Biflagellate sperm may be correlated with internal fertilisation and the need for the sperm to travel through tissue to reach the egg. Yet internal fertilisation is common among invertebrates and the platyhelminth biflagellated spermatozoa are unique among animals.

Another specialisation in most platyhelminths is that yolk is not produced within the egg cell but in separate cells, which are later incorporated into the egg capsule.

The remarkable powers of regeneration of many planarians may also represent specialisation: as little as one-sixteenth of the body can reconstitute a new small animal. Also, in adverse conditions the animal may lose its differentiated structures and shrink, increasing again in size and complexity when the environment is favourable.

6.5.2 Predation and parasitism

Turbellarians are nearly all predators, capturing their food with an eversible pharynx. The complexity of the gut increases in the larger turbellarians: with no transport system, the whole body can be supplied with food only by gut branches.

Parasitism was a fruitful evolutionary direction for platyhelminths (see Box 6.1). Platyhelminths are relatively small and thin and inactive: without any means of transporting food and oxygen round the body they cannot become larger, and their muscle system does not allow much activity. Parasitism is a way of life where these characteristics are not disadvantageous. Energy can be concentrated on reproduction, and the complex hermaphrodite genitalia of free-living platyhelminths, coupled with internal fertilisation, can readily evolve the great fecundity required by parasites. There is a large number of parasitic species with a great variety of life cycles, very often with an intermediate host.

Box 6.1
Parasites

Parasitism is a specialised form of predation. A parasite lives inside, or attached to the outer surface of, the body of a living host of another species, from which it obtains its food. The parasite benefits from the association at the host's expense; typically it harms but does not kill the host and thus become deprived of a home.

The life of a metazoan parasite demands special adaptations: an ectoparasite must be able to attach to the host's outer surface, an endoparasite must penetrate the host, find and perhaps attach itself to a particular tissue and resist the host's defences. A gut parasite must resist the host's digestive enzymes, a blood parasite must adapt its surface antigens to resist the host's immune system. Such adaptations are not made once for all: there is a continuing 'arms race' between parasite and host. This demands continual individual variation on both sides, supplied most rapidly by sexual reproduction. The ubiquity of parasitism and the demands of the arms race may indeed be a powerful reason for the continuation of the apparently wasteful procedures of sexual reproduction.

Internal parasites have little need for organs of locomotion and food capture, but have particular difficulty in completing their life cycle. The young cannot overcrowd the parental host, they must find new hosts: many fail and then die, so that a large number of young must be produced. Typically there is an elaborate life cycle with a series of larvae (perhaps multiplying asexually) as agents of dispersal and host-finding. These larvae frequently find an intermediate host, perhaps one eaten by the primary host, to facilitate transfer.

Parasites have evolved in almost every phylum but are particularly prevalent in Protozoa (which are outside the scope of this book), platyhelminths (see Figure 6.4 for two examples of life cycles) nematodes, acanthocephalans and certain arthropods (see Chapter 13 for special attributes of parasitic crustaceans). Parasites are sometimes described as degenerate because they have lost some of the elaborations of their free-living relatives, but they are more accurately seen as specialised, being highly adapted to their particular way of life. They are very widespread and likely to be important in any ecosystem.

Parasites may even alter the behaviour of their hosts to suit their own requirements: for example, trematode flukes parasiting certain snails cause the host to seek the light, bringing them to open places where predation by birds is much more likely. This brings the flukes to the birds, which are their next hosts, but does not at all benefit the snails. This behaviour change has been attributed to action by the parasite on the host hormones, but that a parasite may directly affect gene expression has been shown in another species of trematode-infested snail. At infection this snail suffers a three-fold increase in production of the messenger RNA (mRNA) which codes for a brain neuropeptide, causing the snail to cease all sexual activity and rapidly increase in size: it can then accommodate a larger number of the parasites.

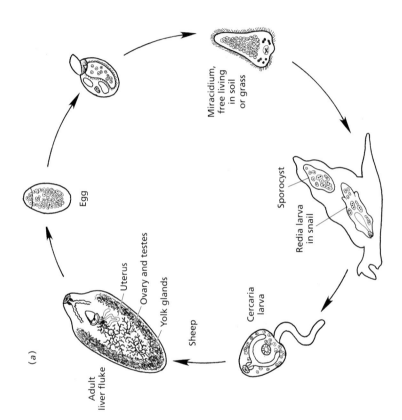

(a)

Adult
liver fluke

Uterus

Ovary and testes

Yolk glands

Sheep

Egg

Miracidium,
free living
in soil
or grass

Sporocyst

Redia larva
in snail

Cercaria
larva

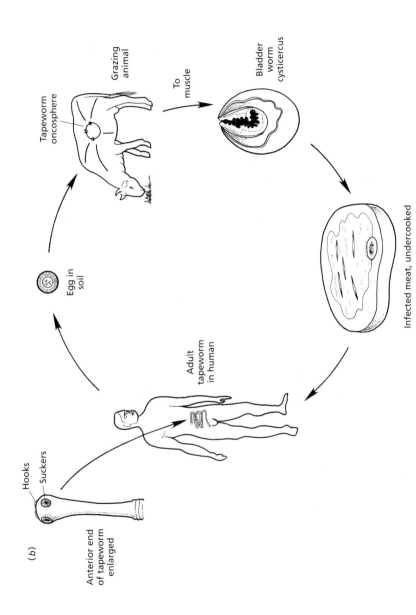

Figure 6.4. The life cycles of two parasitic platyhelminths: (a) a liver fluke, *Fasciola*, a trematode; (b) a tapeworm, *Taenia*, a cestode.

(b)

Hooks
Suckers

Anterior end
of tapeworm
enlarged

Adult
tapeworm
in human

Egg in
soil

Tapeworm
oncosphere

Grazing
animal

To
muscle

Bladder
worm
cysticercus

Infected meat, undercooked

For example, in the liver-fluke *Fasciola* (Figure 6.4*a*), where the adult lives in a sheep, the eggs are shed into damp grass where they hatch into free-swimming larvae (unlike the adult this stage has a ciliated epidermis, and simple eyes). These larvae enter the lung of a snail, form cysts that free larvae which multiply asexually to increase their numbers even more, and the final larvae wait in the grass. If they are eaten by sheep, the cycle is complete. Cestodes such as the human tapeworm *Taenia* (Figure 6.4*b*) are even more modified from their free-living ancestors; a 'head' fastened into the host intestine buds off a series of 'proglottides' that are little more than bags of reproductive organs: there is no gut, food is absorbed all over the surface. Again there is a complex life cycle: eggs pass out into the soil where they develop into a hooked embryo. If eaten by a pig, the embryo makes its way to the pig's muscle and encysts, ready to be eaten by a human.

6.5.3 The epidermis

This is another very specialised structure in Turbellaria. The multiplicity of muscular gland organs is probably very significant. Glands produce complex forms of mucus that protect the animal and, in some way not understood, facilitate both locomotion and attachment to surfaces.

6.5.4 Other systems

In Turbellaria, other systems also show great elaboration in detail. The nervous system is a variable mixture of specialised and unspecialised, advanced and primitive components, again with the greatest variation in the groups that are candidates to be the most primitive Turbellaria. There is a corresponding variety of simple epidermal sensory receptors. Muscle is not well developed, as would be expected in animals where there is so little provision for their re-extension after contraction, yet the ultrastructure of some of the faster muscle is reminiscent of striation (see Box 5.1).

The complexity both of the reproductive system and of the epidermal structures has led some workers to believe that Turbellaria are not primitive animals but are secondarily derived from coelomates by loss of the body cavity, much as among the annelid worms leeches have a reduced coelom, which has been invaded by tissue; this idea requires more evidence.

6.6 How are platyhelminths related to each other?

6.6.1 Relationships within the whole phylum

Parasitic adaptations are universally agreed to be secondary: the common ancestor must have been a free-living flatworm. The ultrastructure of the tegument is unique and is common to all three parasitic classes, strongly suggesting their joint origin from a common ancestor. But do the Turbellaria, and indeed the Platyhelminthes as a whole, form a natural group with a common ancestor? Animals cannot be united by shared primitive characters and union by negative characters is unconvincing. There is not a single shared secondary ('derived') character that unites the three main groups of turbellarians, nor do they share other than primitive characters with the parasitic classes. Turbellaria cannot in any case be what is called a monophyletic group, including an ancestor and all of its descendants (see Chapter 2), since the parasitic classes arose from within the group. Since turbellarians may even be polyphyletic (having no common ancestor), it has been proposed that the Class Turbellaria should be abolished. This suggestion has been put to one side, partly because it is not supported by molecular evidence but mainly because the term 'Turbellaria' is so convenient.

6.6.2 Relationships within the Turbellaria

These are even more controversial; here simplicity is a very uncertain guide. Morphological diversity is greatest in the groups that appear to be the most primitive, which makes it even harder to trace their evolutionary origin. Nor is further evolution easy to trace; for example, in the Rhabditophora the yolk is primitively inside the egg cell, while in many more advanced turbellarians there are separate egg and yolk cells, yet classification according to this character does not accord with classification by the elaboration of the gut. At present these contradictions have not been resolved.

In summary: if the turbellarians have no common ancestor we may be mistaken in regarding platyhelminths as a single phylum. All we can be sure of is that there has been so much convergent evolution among these flatworms that their phylogenetic relationships are very difficult to trace. Different investigators will emphasise different morphological characters, and comparison with the results of molecular methods may be particularly valuable.

6.7 Can platyhelminths represent our ancestors?

Platyhelminths appear to be a collection of primitively simple animals, which may or may not have a common 'flatworm' ancestor, showing a remarkable degree of variation within the constraint of their basic structure. Survival for such animals has perhaps necessitated either the specialisations of parasitism or, in free-living members, the evolution of a large number of small-scale specialisations. The difficulty of identifying the most primitive group of turbellarians reinforces this view. Therefore we should beware of taking either the morphology or the molecular constitution of any modern flatworm to represent the ancestral condition. Further, the whole concept of a flatworm ancestor for all triploblasts is being revised (see Chapter 20).

7

Nemertea

Nemertea (also called Nemertinea or Rhynchocoela) are 'proboscis worms', having a proboscis separate from the gut. They are often called ribbon worms, since they are typically long and thin (Figure 7.1). They are primitively and predominantly marine, and often beautifully coloured.

Nemertines are not so much a minor phylum as a neglected phylum. Over a thousand species are now known, many having been very recently discovered, and more are being found at a rapid rate now that interest in the phylum has arisen. They are here given a separate chapter partly because many aspects of their biology, including their controversial evolutionary relationships, are very unusual and worthy of study, partly also because the author's interest in this phylum has been sharpened by her work on terrestrial and freshwater nemertines.

The phylum shares some basic structural features with the Platyhelminthes. Although the traditional derivation of nemertines from these flatworms is no longer supported, comparison with platyhelminthes is a good approach to the phylum. Consideration of the characters by which nemertines differ from platyhelminths provides both a definition of the distinctive body plan of nemertines and an introduction to some of the most interesting features of the phylum.

7.1 What are the principal groups of nemertines?

See Figure 7.2*a,b*.

- **Anopla:** the proboscis has no stylets: the mouth is below or behind the brain.

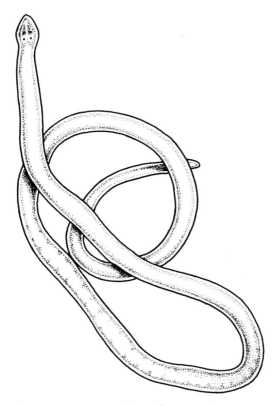

Figure 7.1. External view of a nemertine.

- **Palaeonemertea**, e.g. *Cephalothrix, Tubulanus*. The nerve cords are peripheral and the worms have many simple features.
- **Heteronemertea**, e.g. *Lineus, Cerebratulus*. These are the largest and most muscular nemertines; the main muscle layer is the outer longitudinal one.

- **Enopla**: the proboscis bears stylets: the mouth is anterior, opening with the rhynchocoel.

  - **Hoplonemertea**, e.g. *Paranemertes, Prostoma, Argonemertes*. The central nervous system is inside the body wall muscle: the outer muscle layer is circular.
  - **Bdellonemertea**, e.g. *Malacobdella*. Specialised filter-feeding commensals living in the mantle cavities of bivalve molluscs.

7.2 How do nemertines resemble platyhelminths?

They have three body layers, ectoderm, mesoderm and endoderm enclosing a gut. There is no general body cavity (i.e. they are acoelomate) and they are unsegmented. They have no rigid cuticle, being bounded by a ciliated epidermis, nor other hard skeleton. They move forward by ciliary movement (if they are sufficiently small) and by muscles squeezing on the semi-fluid unspecialised parenchyma. The anterior end has sense organs such as eyes, and there is a simple brain and nervous system where nerve cell bodies and fibres are barely sorted out. Respiratory exchange occurs all over the body surface. There are excretory ducts made from ectodermal intuckings, often with flame cells (see Box 9.1) at the blind endings.

7.3 How do nemertines differ from platyhelminths?

Apart from the absence of platyhelminth specialisations (see Chapter 6) the most pronounced differences are:

1. The proboscis.
2. The gut has two openings, mouth and anus.
3. The presence of a closed blood system.
4. Unique cerebral sensory organs.
5. Simple gonads, regularly repeated along the body.
6. The worms are characteristically very long and thin, and capable of great changes in length.

7.3.1 The proboscis

This lies in the rhynchocoel, a separate fluid-filled cavity dorsal to the gut. Like a coelom it is a space surrounded by mesoderm, but it is not a general body cavity; it contains only the proboscis. Contractions of the rhynchocoel and body wall muscles evert the proboscis and a retractor muscle pulls it back into the worm (Figure 7.3). The proboscis is primarily used in food capture: nemertines are typically active predators. It may also be used, especially in terrestrial nemertines, for rapid escape: it is everted, the tip attaches to the ground and the worm pulls itself forward around the proboscis.

The proboscis may or may not be armed with stylets. On this difference, and on the arrangement of the body wall muscle and lateral nerves, rests the main subdivision of the phylum (see above).

L.S.

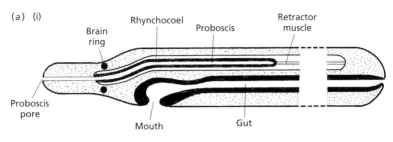

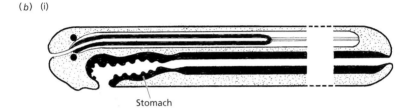

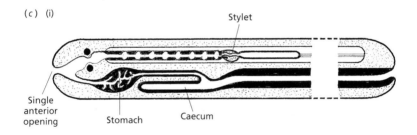

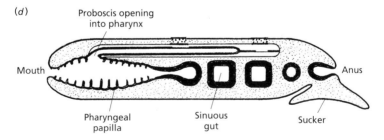

Figure 7.2. Diagrams showing the structure of the four Orders: (a) paleonemertean (i) longitudinal section (L.S.); and (ii) transverse section (T.S.) (showing one common position of the nerve cord); (b) heteronemertean (i) L.S. and (ii) T.S.; (c) monostiliferan hoplonemertean (i) L.S. and (ii) T.S.; (d) bdellonemertean, L.S.

T.S.

(ii)

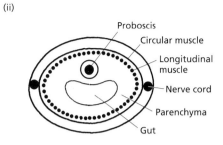

(ii)

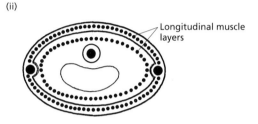

(ii)

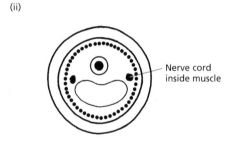

Figure 7.2 (cont.).

7.3.2 The gut

The gut opens at both ends: there is therefore one-way traffic of food and there can be regional differentiation. This not only allows nemertines to become elongated, it also at once demands a blood transport system so that absorbed food can be distributed throughout the worm.

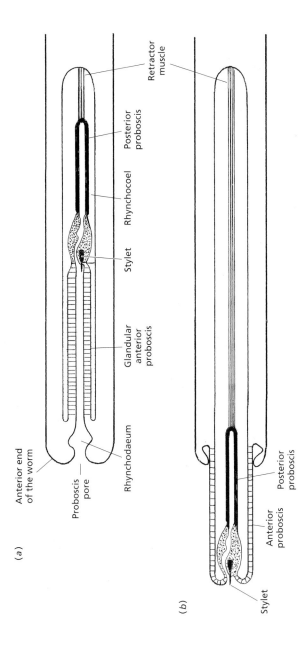

Figure 7.3. Diagrams showing, in dorsal view, the eversion of the hoplonemertean proboscis: (a) proboscis retracted; (b) proboscis everted.

7.3.3 The blood system

The blood system is formed by mesodermal splitting. This method of formation is unique among invertebrates (usually blood is contained in the persisting primary body cavity). The nemertine blood system appears to be a quite separate product of evolution, made necessary by the second opening of the gut. Typically it consists of a loop at the anterior end, with vessels meeting under the brain, and one mid-dorsal and two lateral vessels (often with regular cross-connections) extending posteriorly to join near the anus (Figure 7.4*a*). There may be an elaboration of vessels beside the foregut, and blood is prominent at any site of water transfer: two examples are that the excretory system is often closely associated with blood vessels (Figure 7.4*b*) and that usually the mid-dorsal vessel penetrates the rhynchocoel wall in the brain region and runs for a short distance bathed in the fluid of the rhynchocoel, the proboscis having no other blood supply (Figure 7.4*c,d*).

The blood does not seem to be required for oxygen transport and indeed is situated deeply beneath the body wall musculature. Haemoglobin, if present, is usually in the brain and nervous system.

7.3.4 The cerebral organ

This neuroglandular structure is closely associated with the brain (see Figure 7.5*b*) and is unique to nemertines. It is partly sensory, being concerned with the detection of food, light and the osmotic concentration of the surrounding water. It also regulates the secretion into the blood of acid mucopolysaccharides from which mucus is made. We need more experimental work concerning the function of this unusual organ.

7.3.5 The gonads

Nemertine gonads are simple and very numerous: they are regularly repeated between lobes of the gut along the length of the worm, giving these worms an almost segmented appearance. Most nemertines have separate sexes but some species (including all those known from fresh water) are hermaphroditic.

Most commonly eggs and sperm from different animals are shed from the many gonads and external fertilisation occurs in the sea. The egg may hatch into a planktonic larva or the young may develop directly within the egg. Sexual reproduction predominates, but some species have great regenerative powers that can lead to asexual reproduction.

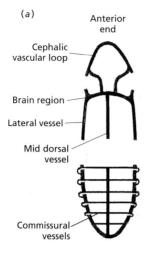

(a)

Anterior end

Cephalic vascular loop

Brain region

Lateral vessel

Mid dorsal vessel

Commissural vessels

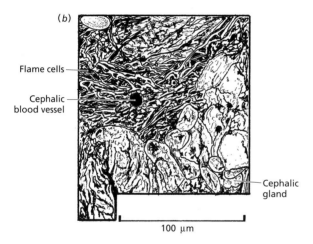

(b)

Flame cells

Cephalic blood vessel

Cephalic gland

100 μm

Figure 7.4. The vascular system of a hoplonemertean: (a) plan of blood vessels (one common version); (b) part of an anterior transverse section (T.S.) of *Geonemertes pelaensis*, a terrestrial nemertine, showing the accumulation of flame cells round a cephalic blood vessel;

(c)

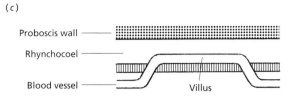

Proboscis wall

Rhynchocoel

Blood vessel

Villus

(d)

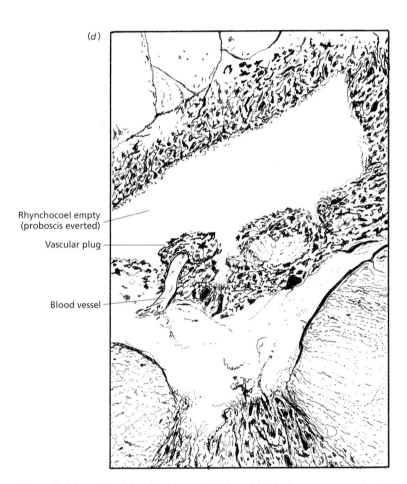

Rhynchocoel empty
(proboscis everted)

Vascular plug

Blood vessel

Figure 7.4 (*cont.*). (*c*) a rhynchocoelic villus; (*d*) T.S. *Argonemertes dendyi* , a terrestrial nemertine, showing two circular vascular plugs in place of a villus; (*b*) and (*d*) drawn from photomicrographs.

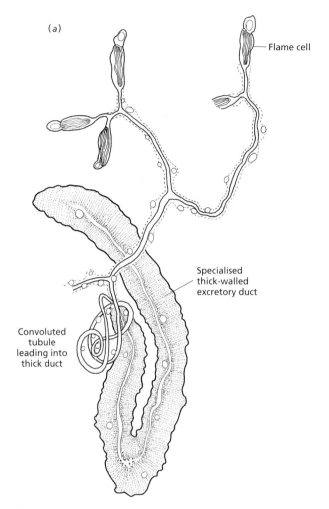

Figure 7.5. *Argonemertes dendyi*: (a) one protonephridium;

7.3.6 Length and extensibility

Nemertines are often very long and thin, and very extensible. Some (e.g. terrestrial and interstitial species) are only a few millimetres long, but some in the sea are very long indeed, for example *Lineus longissimus* can reach 30 metres, while remaining only a few millimetres in diameter. What is the advantage of so great an extension of one animal? There will be a usefully large number of gonads, but why attach them all to one feeding head? The answer probably

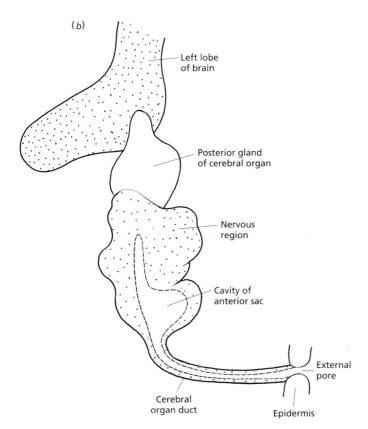

(b)

Left lobe
of brain

Posterior gland
of cerebral organ

Nervous
region

Cavity of
anterior sac

External
pore

Cerebral
organ duct

Epidermis

Figure 7.5 (*cont.*). (*b*) dorsal view of the left cerebral organ.

lies in the nature of the epidermis. It is deeply folded with microvilli and many mitochondria, and in some species experiments with radioactively labelled glucose and amino acids have shown that the epidermis can absorb dissolved organic food as a supplement to normal feeding. Extension in length is therefore a structural economy. A long worm might be more vulnerable to predators, but the epidermis secretes deterrent toxins. A crab taking a nemertine may hold it in one claw and wipe it clean with the other before eating it: not all predators are so wary.

Nemertines are not only often long, they are capable of unusual changes in length: by a factor of 10 in *Lineus*. The lateral nerve includes a longitudinal muscle fibre that assists these changes. Proboscis eversion can involve a 50-fold change in length of the proboscis retractor muscle: the proboscis can be shot out (by fluid pressure caused by muscle contraction) and pulled back (by the retractor muscle) very fast. This combination of great change in length with rapidity of contraction is a most unusual property for any muscle. There has been no recent experimental work on nemertine muscle.

Could there be a connection between contractility and the amount of muscle in the body? 'Hoop stress' to burst a cylinder sideways can be shown to be twice the longitudinal bursting stress (frying sausages burst along their length; they do not break down into little rings). A worm should therefore have twice as much circular as longitudinal muscle in cross-section. Nemertines do not; nor (unlike earthworms) do they use the wall of a burrow for support. Is there some correlation here with very variable contraction?

Other nemertines such as *Cerebratulus* are much thicker and vary little in length. They can swim strongly by vertical undulations passing along the body, which is stiffened by dorsoventral muscle contraction and so does not change in length: leeches do the same thing. Remarkably, if a *Cerebratulus* is suspended half in air and half in water, the part in water will swim while the part in air carries out earthworm-style peristalsis. Clearly there can be very little whole-body coordination of muscle contraction: nemertine muscle and its control should repay further study.

7.4 What diversity exists among nemertines?

Having seen that these worms have many unexpected features, let us look at what natural selection has produced within the phylum:

7.4.1 Morphological diversity

At first sight this is a rather uniform phylum, but there is great and seemingly arbitrary variation on a small scale, sometimes unexpectedly combined in different animals; for example, a curious 'wickerwork' construction of the rhynchocoelic muscle has evolved several times, and one species in the otherwise unarmed Class Heteronemertea has many small stylets on the proboscis. Occasional species have characters quite unexpected for the phylum: one has ganglia along a dorsal nerve, some have striated muscle and one tiny species living between sand grains shows external annulations, rather like an earthworm. There has clearly been a great deal of convergent evolution. Even the simple classification given above is controversial: emphasis on different characters produces different hypotheses about evolutionary relationships. On any scheme, some elaborate and apparently stable characters must have evolved more than once.

7.4.2 Feeding methods

These at once illustrate the theme of much small scale diversity superimposed on apparent uniformity. Nemertines characteristically feed as carnivores: they may be ambush predators or may actively hunt their prey, which commonly consists of crustaceans or polychaete (marine annelid) worms. The proboscis is everted rapidly and accurately and wraps around the prey, which in unarmed species is swallowed whole. Armed nemertines paralyse or kill their prey by piercing it with stylets and secreting toxins. Yet nemertines do not always rely on the proboscis: some absorb organic food in solution all over the surface, for example *Lineus* (see above) and the larval stages of *Carcinonemertes*, an egg predator living on crabs; its adult stages do use the proboscis (they will only evert it on a surface curved like an egg). *Malacobdella* is a commensal (i.e. it shares the food of another animal with which it lives). Here the proboscis is less important and neither mucus nor toxin is secreted. The pharynx expands and the papillae projecting from its inner surface interlock, making a net that entraps food particles. Compressed, the water then passes backward, propelled by cilia at the base of the papillae. This is feeding like a flamingo or a baleen whale.

7.4.3 Ecology

Ecological diversity is considerable. Nemertines are found in all marine habitats from the deep seas to the top of the shore: they are common in both tropical and polar seas. They may live as commensals with tunicates or anemones or as egg predators under the carapaces of crabs. About a dozen species have colonised land, living in damp cool places, for example under rotting logs. Terrestrial adaptations have arisen several times separately, in species from different hoplonemertean families, showing strong convergence; for example, all have great development of the excretory system (Figure 7.5*a*, one of the many thousand protonephridia in a terrestrial nemertine some 10mm long) that removes excess water entering when they are flooded, and extensive mucus production that protects them against desiccation. Rather more species are found in fresh water, again showing great convergence, belonging to both heteronemertean and hoplonemertean families. There appear to have been two routes for colonisation of fresh water, one by way of estuaries and the other overland. See Box 7.1 for a general discussion of life in the sea, in fresh water and on land.

7.5 How are nemertines related to other phyla?

No certain fossil evidence exists. Early development (in most but not all species) is by spiral cleavage (see Box 5.3) and the mouth forms before the anus, thus placing nemertines among 'protostome' invertebrates such as platyhelminths, annelids and molluscs. The traditional derivation from platyhelminth stock has recently been questioned, and molecular methods place nemertines closer to coelomates. The rhynchocoel can be seen as a restricted coelom confined to one part of the animal, and blood vessels develop by splitting of mesoderm in a manner somewhat resembling protostome coelom formation. There has even been a suggestion of coelomate ancestry for nemertines, postulating that coelom reduction gave rise to the whole phylum. Production of 'coelomic' cavities is, however, known to have occurred many times in evolution. Regression followed by radiation into a phylum with more than 1000 species seems very improbable indeed. The conclusion is that nemertines have arisen from an early acoelomate stock, but may well be closer to present-day coelomates than to present-day platyhelminths, which are distinctly specialised animals (see Chapter 6).

There is no convincing evidence that any other phylum has arisen from early nemertines. Suggestions have been made that they gave rise to other phyla, even to chordates and hence to vertebrates, but these ideas are not accepted by most morphologists, nor are they in line with molecular evidence. Convergent evolution of certain characters seems a much more probable explanation for any resemblances.

Box 7.1
Sea, fresh water and land

Sea

Animal life began in the sea. The marine habitat is favourable, being a large body of water fairly constant in quantity, movement and salt concentration. It is buoyant, relatively rich in potential food and it enables animals to disperse. Many marine invertebrates shed their eggs and sperm directly into the sea and fertilisation is external. The eggs contain little yolk and hatch early, usually into larvae feeding in the plankton (surface waters). Above all, the body fluids of marine invertebrates are similar in salt concentration to that of the sea in which they live, and there is no osmotic problem.

Units and definitions

The osmotic pressure of a solute is usually measured by the **depression of the freezing point** (Δ) caused by the solutes. Pure water freezes at 0.00 °C; sea water (with salt 35 parts per 1000) at -1.86 °C. **Osmoles** measure the osmotic effect of a solute whose molecular weight in grams is dissolved in a litre of sea water. For sea water the value is 1 osmole ($=$ 1000 milliosmoles, mosm) and this provides a useful standard for comparison. Most marine invertebrates are **osmoconformers**, with body surfaces fully permeable to water, i.e. the body fluid concentration is close to one osmole.

— **Osmotic regulation** by contrast is the maintenance of the internal body fluids at a concentration different from that of the environment.
— **Euryhaline** animals are those which can tolerate a wide range of salinities, usually by a combination of osmoconforming and osmoregulating processes.
— **Stenohaline** animals are those (like most marine invertebrates incapable of osmoregulation) that are restricted to a narrow salinity range.
— **Ionic regulation**, selection of particular ions at concentrations different from those in the sea does, however, occur even in the simplest marine animals. A jellyfish floats easily because it contains, relative to the surrounding sea, fewer heavy ions (magnesium, sulphate) and more lighter ions (sodium, chloride). A marine lobster (*Homarus* sp.) maintains concentrations of magnesium ions at 14% and sodium at 111% sea water, but it is incapable of osmoregulation.

Fresh water

Colonisation of fresh water from the sea raises problems. The medium is much more variable in many ways: it may overheat or freeze, it may flow very fast, it

Box 7.1 (*continued*)

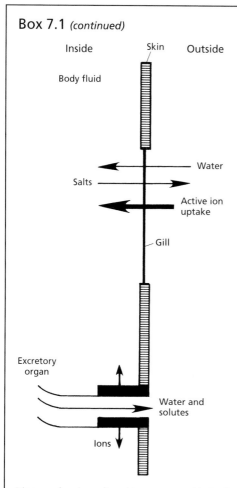

Inside   Skin   Outside

Body fluid

Water

Salts

Active ion uptake

Gill

Excretory organ

Water and solutes

Ions

Diagram showing salt and ion movement in freshwater animals. Note, active ion reabsorption in the excretory organ occurs only in some of the freshwater animals.

may dry up, and above all it is much more dilute (5–10 mosm) than internal fluids can be. Water will therefore tend to enter by osmosis, and salts (ions) to leave by diffusion. If these movements were unchecked, animals would swell up and burst or all the salts would leak away. Small animals, with relatively more surface area in proportion to their volume, have a worse problem than larger ones.

Long-term adaptations to life in freshwater include:

1. **Reduction of the permeability of the outer surface**: some permeability must remain, if only in the gills, to allow intake of oxygen for respiration and removal of carbon dioxide.

2. **Reduction of the internal concentration** to reduce the difference between inside and outside. The body fluids of freshwater animals are almost always more dilute than those of their marine relatives (perhaps 100–300 mosm.) but they cannot be as dilute as the medium.
3. **Tolerance of fluctuations** in the salinity of the body fluids is characteristic of many estuarine osmoconformers, for example annelids, molluscs and crustaceans. Estuarine animals need to be euryhaline. One adaptation is to retain amino acids within the cells: when the body fluids are diluted by osmotic influx of water, the cells release the amino acids, becoming temporarily as dilute as the blood and therefore not subject to entry of water.

The above adaptations are insufficient by themselves to support life in fresh water: osmoregulation, an energy-consuming process, is required. Water is never transported actively: ions are moved, and water follows by osmosis:

1. **Active uptake of ions** at the gills or over the whole body surface occurs in all freshwater animals. Usually sodium or chloride ions, or both, are taken in, compensating for the outward diffusion of salts. Entering water is removed by the excretory organ.
2. **Active reabsorption of ions** in the excretory organ, with the production of urine more dilute than the blood (hypotonic), may reinforce ion uptake at the surface. This occurs in many but by no means all freshwater animals.

Reproductive adaptations

Sperm do not survive well in fresh water and external fertilisation is rare. More freshwater animals are hermaphrodites. The eggs typically contain much yolk and most develop directly: a fragile larva, small with a relatively large surface area, would be very vulnerable to osmosis/salt loss and liable to be swept downstream.

Fresh water was colonised only in part by the aquatic route up estuaries: many freshwater animals (nearly all of the molluscs and arthropods) evolved from terrestrial forms secondarily returning to water.

Land

Many so-called terrestrial invertebrates live in soil (e.g. earthworms, planarians, nemertines, nematodes) or other moist places, and remain physiologically aquatic.

Problems for animals colonising land

These include:

1. Avoidance of desiccation.
2. Tolerance or avoidance of temperature fluctuations.
3. Support and locomotion of the body in air.

Box 7.1 *(continued)*

Not surprisingly, most truly terrestrial invertebrates are arthropods or molluscs. When surrounded by dry air rather than water animals must:

4. Respire using lungs rather than gills (see Box 9.2)
5. Excrete with minimal water loss (see Box 9.1)
6. Perceive different environmental stimuli, often with different sense organs.

Routes by which land was colonised

Many small animals (about 2 mm long) came by the interstitial route, evolving from ancestors living in the spaces between grains of sand or other sediments. Most terrestrial animals, however, came to land either by the aquatic route from fresh water (e.g. vertebrates) or by the overland route from the top of the seashore (e.g. most arthropods and many molluscs). The routes may be identified by differences between present-day animals:

1. Animals once adapted to fresh water have body fluids more dilute than the sea and this difference will be maintained in their further evolution (e.g. the body fluids of slugs and snails range from 97–231 mosm).
2. Terrestrial animals whose ancestors took the overland route are likely to have body fluids nearer to marine concentrations (e.g. 700 mosm in the woodlouse *Porcellio*). Animals evolving by way of the top of the seashore are also likely to tolerate some evaporative water loss.

8

Nematoda

Nematodes, or 'roundworms', are totally different from platyhelminths and nemertines. A nematode is characterised by an extremely tough thick cuticle round the outside and a very high hydrostatic pressure within. They all look very similar (Figure 8.1a) and given the pressure it is hard to see how any other shape could be maintained, yet there are perhaps a million species. The phylum is unusually ubiquitous; nematodes are free-living in marine, freshwater and land habitats and parasitic in animals and plants. They are clearly of great economic importance as they exist in extraordinary numbers and play a significant role in the total matter and energy cycle of the biosphere. R. Buchsbaum (*Animals Without Backbones*, 1938) wrote, 'If all the matter in the universe except the nematodes were swept away, our world would still be dimly recognisable, and if, as disembodied spirits, we could then investigate it, we should find its mountains, hills, vales, rivers, lakes and oceans. The location of towns would be decipherable . . . Trees would stand in ghostly rows . . .', etc. We are told that four and a half million individual nematodes were found in one square metre of Dutch marine mud, and 90 000 in one rotten apple.

What makes these worms so remarkably widespread and numerous? The cuticle is clearly the answer, and it is argued below that the combination of tough cuticle and high internal pressure dictates most of their (often unique) characteristics. Their development is also discussed: like their morphology it shows unusual uniformity in that a nematode has a fixed number of large cells of predetermined fate. This offers opportunities to biologists for study of the control of development by genes, with results applicable to other animals, including ourselves. Nematodes have been very harmful to humans in causing disease and in damaging agriculture, but study of their development has given us important medical tools.

8.1 What are the distinctive characters of nematodes?

Nematodes are slender worms, circular in cross-section and ranging in length from 200 μm to 40 cm (in *Ascaris*, a large intestinal human parasite) or even 9 metres in a parasite of whales. The worms are triploblastic, unsegmented and enclosed in a tough but flexible cuticle, under which is a layer of muscle, longitudinal only. There is no blood or other circulatory system. The body cavity is termed a pseudocoel, since it is probably derived directly from the blastocoel (see Box 5.2) with no inner lining of mesoderm. It contains the many tubes of the reproductive system and also the gut, which opens at both ends of the worm. The general structure (Figure 8.1*b,c*) is a series of tubes within a tube.

8.2 How are these characters related to the cuticle and fluid pressure?

8.2.1 The cuticle

This organ is a many-layered structure toughened by cross-linked protein chains and incorporating a collagen lattice which enables movement (see Chapter 5). Figure 8.1*d* indicates the structure in *Ascaris*, with nine layers in three zones. The cuticle is laid down by an underlying 'hypodermis' (the equivalent of the outer epidermis of other worms), which is syncitial. The cuticle is extremely resistant both mechanically and chemically: it enables nematodes to survive in deserts and polar regions and to withstand the digestive enzymes of host animals. It is undoubtedly fundamental to the success of nematodes as parasites and it has been very hard to develop pesticides effective against them.

8.2.2 The body cavity

The body cavity contains fluid at pressure that may be an order of magnitude higher than that of an earthworm, which operates at 3–30 cm water (0.28–2.8 kN per square metre). The figures for *Ascaris* are 70–400 cm water (6.6–37.6 kN per square metre).

8.2.3 The locomotor system

The locomotor system depends entirely upon muscle working against the cuticle and the internal pressure. There are no cilia: they could beat neither on the outside nor on the inside. Movement is achieved by contraction of four blocks of longitudinal muscle, which (most unusually) send processes to the nerves rather than receiving nerve endings. The muscle is further unusual in having contractile and non-contractile regions (the contractile parts are obliquely striated with a pattern of thick myosin and thin actin filaments, as shown in Box 5.1). The cells are very large: a single muscle cell may be 1 cm in length. Circular muscles are absent: they could not contract under the cuticle and are in any case redundant, because longitudinal contraction on one side of the worm is opposed by pressure making the cuticle bulge on the other side, guided by the collagen lattice, without overall change in shape. Since there is no circular muscle, the lattice angle must at all times be greater than 55° (see Chapter 5). Each half of a piece of the worm alternately contracts and relaxes, throwing the animal into lateral waves. These waves are propagated backward along the worm, making it move forward with the undulating progression that is so characteristic of nematodes (see Figure 5.1c). Locomotion is only possible when there is soil or tissue for the body to press against: a freely floating nematode will thrash about ineffectively.

Coordination in such a system is achieved largely by hydraulics. There is a minimal nervous system, with longitudinal nerve cords and an anterior nerve ring in place of a brain. Sensory structures may be present at either end as small projecting tactile papillae or chemosensory pits.

8.2.4 Other systems

Other body systems are similarly governed by the cuticle and high internal pressure. Feeding is by a mouth pump, sometimes assisted by teeth or stylets, drawing in semi-liquid food. A strong muscle closes the anus to keep the food inside. Respiration is often largely anaerobic, based on partial breakdown of stored glycogen (animal starch). In a classic early experiment, nematodes were buried in anoxic mud from the River Clyde in Scotland and found still to be active after 35 days. Where oxygen is provided in the environment, it diffuses in over the body wall and combines with the haemoglobin (which has a particularly high affinity for oxygen).

Excretion has been attributed to two large 'renette' cells that may be elongated into H-shaped tubes: these single cells run the length of the body.

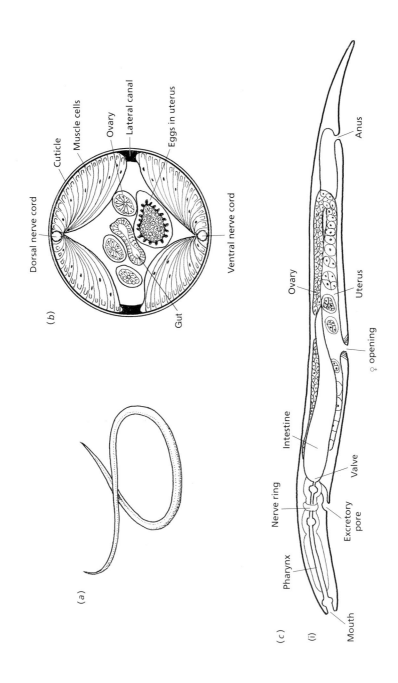

(a)

(b)

Dorsal nerve cord

Cuticle
Muscle cells
Ovary
Lateral canal
Eggs in uterus

Ventral nerve cord

Gut

(c)

(i)

Nerve ring
Pharynx
Mouth

Excretory pore
Valve

Intestine
Ovary
Uterus
♀ opening

Anus

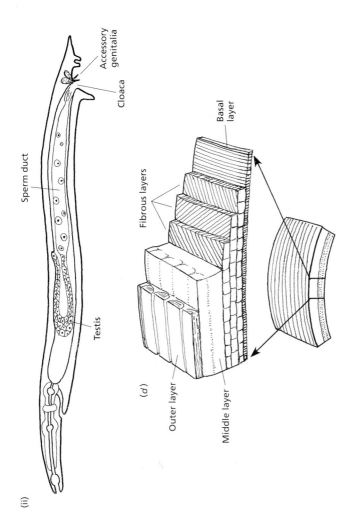

Figure 8.1. The structure of nematodes: (a) external view; (b) transverse section based on *Ascaris*; (c) longitudinal section of (i) female and (ii) male based on *Rhabditis*; (d) nematode cuticle.

However, experiments showed that urea is voided by the anus rather than by the tube's pore. The ultrastructure of the tube resembles that of the contractile vacuole of *Amoeba*, suggesting an osmoregulatory function (see Box 9.1). Note that while function cannot safely be inferred from large-scale structure, at the smallest level structure and function are so closely associated that such inference becomes respectable.

8.2.5 Reproduction and life history

Most of the space inside a nematode is filled by the reproductive system. The sexes are usually separate, fertilisation is internal. Again the high internal pressure makes problems: the oviducts have to be kept open with spicules supplied by the male penis. The sperm are amoeboid, as in various other animals with internal fertilisation; in nematodes, probably flagella could not beat against the pressure (note that we know so little about amoeboid movement that we perceive no problem about that). The numerous 'eggs' laid are in fact zygotes or early embryos: they hatch as sexually immature miniature adults.

The growth of the young nematode is punctuated by moults, although the cuticle does allow some increase in size. Consequently the different stages are described as larvae, although their similarity to the adult makes 'juvenile' a better term. In parasites the third of these four stages is characteristically a wandering stage, when a new host or part of a host is colonised. A large volume of reproduction is most important for a parasite (see Box 6.1) and an elaborate life history may facilitate transfer between hosts.

8.3 How is the phylum subdivided?

The unusual degree of morphological uniformity makes classification extremely difficult. There is no help from fossils, which are few and recent. Primitively free-living and probably marine, nematodes (unlike platyhelminths) seem to have evolved parasitism many times independently, with far-reaching convergence making it even harder to trace their evolutionary relationships. Many families have been named and recently molecular methods have made much progress, but identification of nematodes remains a task for experts.

8.4 How are nematodes related to other animals?

Morphological evidence has failed to answer this question, and molecular comparisons have been hampered by the unusually rapid rate of nucleotide substitutions in nematode ribosomal DNA. However, emerging evidence rather surprisingly places the phylum among other animals with moulted cuticles, including arthropods (see Chapter 20).

8.5 Why are nematodes useful for developmental studies?

Reproduction is sexual: budding and regeneration cannot occur. The early development of nematodes has long been known to be very unusual. In the late nineteenth century, Theodor Boveri studied a species of *Ascaris* with two chromosomes and observed 'chromosome diminution', the loss of parts of them in early cell divisions. We now know that at the 16 cell stage a full set of genes is preserved in only two cells, one of which will give rise to the germ cells. In Boveri's time this procedure was not seen as particularly odd, and emphasis fell on the unique 'nematode T', a temporary configuration of the first four cells (see Figure 19.3). Modern study has focused on the remarkably small number of cells in an adult nematode and the fixity of their developmental origin. From the start there is an invariable sequence of cell divisions, each determining the nature and future position of the daughter cells. Early in development, cell divisions cease, so that for a particular species there is a fixed number of cells both in a given tissue and in the whole nematode (about 1000 somatic cells on average). Further enlargement of the worm is due solely to increase in size of its cells, which may become remarkably large in a large adult. Even more remarkably, the fate of each cell depends upon its lineage; most developing animals override information from cell lineage with regulating messages between neighbouring cells. It is this peculiarity that makes nematodes unusually accessible to experimental study of development and its genetic control.

   As so often in Biology, the above generalisations have to be qualified because natural selection is at work. The picture of 'determinate' development with lineage deciding the fate of cells is correct and important, and carried to a greater extreme in nematodes than in any other animals known, but there are exceptions. Some tissues (the hypodermis and the gut) of some species do have variable cell numbers, and regulatory action between cells does occur at various stages in development).

(a)

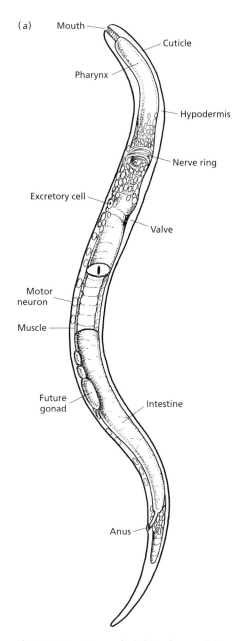

Mouth

Cuticle

Pharynx

Hypodermis

Nerve ring

Excretory cell

Valve

Motor
neuron

Muscle

Future
gonad

Intestine

Anus

Figure 8.2. *Caenorhabditis elegans*: (a) young specimen, with internal structure seen by transparency;

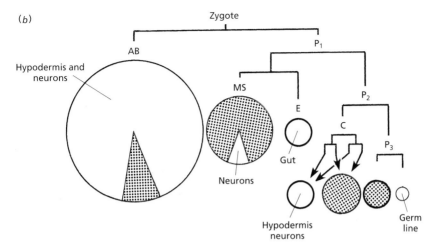

Figure 8.2 (*cont*.).   (*b*) the first divisions of the zygote, indicating how the future fate of the daughter cells is progressively determined at each division. P_{1-3} represents the germ line to future gonads and gametes. Muscle develops from all four shaded areas. Note multiple origin of neurons also.

Caenorhabditis elegans (see Figure 8.2*a*) was chosen over 30 years ago for detailed study. A free-living soil nematode 1 mm long, it can be cultured in the laboratory and at 20 °C it completes its life cycle in 3.5 days. Adults are either hermaphrodites or males. Since the worm is transparent, observation of cells in life can give continuity and certainty and can be combined with videotape recording. The complete cell lineage is known: the earliest steps in the sequence are shown in Figure 8.2*b*. Some interesting findings are:

1. At hatching the hermaphrodite has 550 non-gonadal nuclei, increasing to 810 at maturity. The mature male has 959. Cell numbers for different tissues are: nerve 302, muscle 122 and intestine only 34.
2. Cells of a kind do not always arise from the same germ layer (i.e. ectoderm, mesoderm, endoderm). Identical muscle cells arise from quite distant positions; nerve cells are formed at three separate stages in the lineage (see Figure 8.2*b*).
3. Cell death is programmed to occur at certain stages in the lineage.
4. Regulative cell interaction occurs, for example, in the formation of the vulva, a non-essential structure because in its absence the hermaphrodite can fertilise itself. Two adjacent cells in the female gonad interact so that one of them becomes the uterine precursor and the other the 'anchor cell'; this cell first initiates the cell lineage that makes a vulva and then

induces the ventral hypodermis to make it an external opening. Recent work revealed further regulative events stages, even between the first two cells (see Chapter 19), but determinate development strongly predominates.

Genetic analysis is based on developmental mutants, and is helped by the ability of the hermaphrodite to be either self- or cross-fertilised. Cell lineage defects in mutants can be precisely described, and alterations can be understood in terms of altered developmental decisions in individual cells. Mutations may result in discrete transformations of cell fate at any point in a lineage, or may switch development between lineages. One gene, for example, can alter the rate of ageing throughout the animal.

Uniquely among invertebrates, the entire genetic sequence of *C. elegans* (which has some 18 000 genes) is now fully known. The genetic control of the developmental fate of each cell is largely understood, in terms of a hierarchy of cell decisions governed by particular genes. Many of the largest gene families make regulatory proteins that determine the effects of other genes.

The extent to which genes have been conserved throughout evolution is the most surprising and important discovery to emerge from recent work on nematodes, fruitflies and mammals. *Caenorhabditis elegans* has equivalents of about half of known human genes, some of which offer medical opportunities; for example, the 'ageing gene' mentioned above has a human counterpart that makes a receptor for insulin and there are nematode genes controlling programmed cell death that may become important in treating cancer (where cells fail to die). The uniquely full understanding of the genetic regulation of development in nematodes can be applied to investigating other animals, including ourselves.

8.6 Conclusion

Nematodes are exceptionally uniform and widespread, combining uniformity of structure with variety of habitat. It has been said that there appears to be only one nematode, the model coming in various sizes and with a great variety of life histories. Box 6.1 gives an outline indication of their importance and diversity as parasites. Their success and their morphological uniformity can both be attributed to the unique structure and properties of the cuticle. Their developmental uniformity has been particularly useful to biological and medical research, as the fixed hierarchy of cell decisions can be used to elucidate general features of the genetic control of development.

9

Annelida

Most of the 12000 known species of annelid worms live in the sea, crawling under rocks on the shore and sea bottom, freely swimming or seeking protection from predators in burrows or tubes. Freshwater and terrestrial annelid radiation has been small except for the earthworms, which are outstandingly successful, and the leeches, which are widespread specialised suctorial predators. Annelids vary in size from 1 mm between sand grains to 3 metres in some Australasian earthworms.

Annelid worms are coelomates with metameric segmentation. The annelid coelom is a large fluid-filled body cavity surrounded by mesoderm, providing an efficient hydrostatic skeleton. Metamerism is the serial repetition of similar parts along the length of an animal, manifested primarily in the separation of the mesoderm into segmental blocks of muscle. In annelids there are usually internal partitions (septa) between the segments. Both the coelom and metamerism improve the effectiveness of muscle contractions, so that active locomotion can be much faster than in acoelomates. At the same time these advances demand greater complexity: when an extensive coelom separates the inner and outer tissues, a transport system is required and more elaborate respiratory and excretory organs may need to develop. This complexity then allows more structural differentiation and an increase in size.

The following account first defines and subdivides the phylum, then describes locomotion to show the advantages of coelomate and metameric organisation, introduces the blood, respiratory and excretory systems to illustrate the increased complexity required by a coelomate, and refers to feeding and reproduction to indicate the diversity within the phylum.

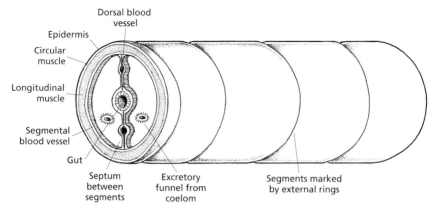

Dorsal blood
vessel
Epidermis
Circular
muscle
Longitudinal
muscle
Segmental
blood vessel
Gut
Septum
between
segments
Excretory
funnel from
coelom
Segments marked
by external rings

Figure 9.1. Diagram of basic annelid structure, with end cut transversely. Alternate sections will have coelom in place of the septum.

9.1 What is an annelid?

An annelid is a worm with three cell layers, a gut with a mouth and an anus, and a body wall with both circular and longitudinal muscle. The coelomic body cavity is formed by the splitting of embryonic mesoderm. The outer epidermis is covered by a thin cuticle, typically bearing chitinous bristles ('chaetae' or 'setae'). Metameric segmentation, always shown in the muscular and nervous systems, is characteristically also evident externally (Figure 9.1). The nervous system has a supra-oesophageal ganglion (i.e. group of the main bodies of nerve cells), which is called a brain although it may be little more than a sensory relay, and a ventral nerve cord bearing segmental ganglia giving off segmental nerves. There is a closed blood system with the blood moving forward in the dorsal longitudinal vessel. Segmental ducts between the coelom and the outside are used for excretion (see Box 9.1) and reproduction.

9.2 What annelids are there and how are they related ?

The three main classes are:

- **Polychaeta.** Typically marine annelids, with 'parapodia', i.e. paired lobes on the sides of most segments forming paddle-like appendages and bearing bundles of chaetae (Figure 9.2*a,b*).
- **Oligochaeta.** Mainly freshwater or terrestrial annelids, without parapodia and with few chaetae (Figure 9.2*c,d*). They are characterised by a saddle-like 'clitellum' an epidermal thickening that secretes a cocoon.

Box 9.1
Excretion

Narrowly defined, excretion is the removal of the waste products of metabolic activity. Since excretory organs are often involved in osmoregulation (in fresh water) or water conservation (on land), the definition may be extended to include more general maintenance of a fairly constant internal environment, by adjustment of the water and ion content of the body fluids. See Box 7.1, relevant throughout to excretion.

Excretory products

Other than water and ions, excretory products include:

— **Carbon dioxide** (CO_2), a product of respiration, in all animals.
— **Ammonia**, NH_3, ion NH_4^+, the primary product of deamination of amino – acids. It is toxic and requires much water for its removal.
— **Urea**, $CO(NH_2)_2$, less toxic and requiring less water loss; it is a very common end-point of nitrogenous metabolism, produced by condensing one molecule of CO_2 with two molecules of ammonia.
— **Uric acid**, $C_5O_3N_4H_4$, the end-point of purine metabolism; it is a more complex molecule, more expensive to make. It can, however, be excreted semi-solid with very little water loss, or stored harmlessly within animals or eggs. It is commonly produced by terrestrial animals, for example insects and snails (spiders use another purine, guanine).

   Most animals excrete a mixture of these substances (and others) but one nitrogenous end-point usually predominates. Ammonia, for example, is 80% of the total in the annelid *Aphrodite*, 60% in the crayfish *Astacus*, 67% in the cuttlefish *Sepia* and 39% in the starfish *Asterias* (where most of the nitrogenous waste is in the form of amino acids). In the land pulmonate snail *Helix*, uric acid constitutes 70% of nitrogenous waste, while in the lower shore periwinkle *Littorina littorea* it is only 0.8–1.2%. Woodlice are exceptional among terrestrial animals in being able to puff out gaseous ammonia.

Excretory organs

Sponges, cnidaria and echinoderms have no excretory organs. Many other marine invertebrates may give out ammonia as well as carbon dioxide over the body surface or the gills, as well as through excretory organs. Apart from unusual structures in nematodes (see Chapter 8) and many arthropods (see Chapter 15 for insect Malpighian tubules), excretory organs are nephridia or coelomoducts:

Box 9.1 *(continued)*

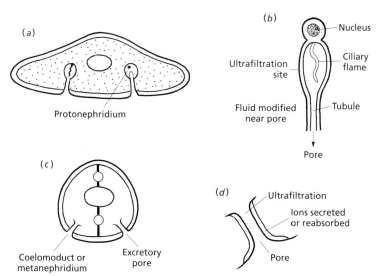

Transverse section (T.S.) of acoelomate with two protonephridia (the left one is shown as a flame cell, and both are disproportionately large); (*b*) flame cell; (*c*), (*d*) T.S. coelomate with two excretory organs (which might be either metanephridia or coelomoducts).

— **Nephridia** are invaginated ectodermal tubes. Those with closed ends are **protonephridia**, flame cells or similar structures (see Fig) where beating of cilia inside the end cell draws water and solutes into a tube leading to the outside. Protonephridia occur in platyhelminths, nemertines, priapulids, gastrotrichs and many larval and a few adult polychaetes.

— **Metanephridia** open by a ciliated funnel from the coelom. They occur in most polychaetes and in oligochaetes and leeches (see Chapter 9).

— **Coelomoducts** are mesodermal ducts from the coelom to the outside. They may be excretory ducts or gonoducts or the two combined, in molluscs and aquatic arthropods. In most molluscs, the pericardium (epithelium round the heart) is the site of filtration of waste from blood in the haemocoel to the much reduced coelom and its ducts.

How excretory organs work

Two processes are involved:

— **Ultrafiltration** Pressure forces water and small dissolved molecules through a semipermeable membrane that holds back large molecules such as proteins. The filtrate in the excretory tubule is modified as it passes along selected substances being added or reabsorbed, by:

> — **Active transport**, usually of ions (never of water). While normally it is superimposed on the filtration process, where there is little blood pressure (as in the insect haemocoel), active transport may also initiate excretion.
> See further discussion of excretion in the chapters on arthropods.

– **Hirudinea** (leeches). Leeches have a clitellum, anterior and posterior suckers and no chaetae (Figure 9.2*e,f*). They are specialised predators, often ectoparasites, in fresh water, on land or sometimes in the sea. Oligochaeta and Hirudinea may be united as **Clitellata**.

The relationship between the three main classes of annelid is clear only in that the oligochaetes are a natural group from which the Hirudinea are a united specialised offshoot. The origin of the Polychaeta is uncertain. 'Archiannelida', a collection of small and therefore simple annelids, were once thought to be ancestral but are now realised to be neither united nor primitive: they are aberrant polychaetes. Relationships among the many orders of polychaetes are very hard to determine: clearly the protection of burrows and tubes was achieved many times separately by different ancestors. Opinions differ about whether oligochaetes arose within the polychaetes or directly from a common burrowing annelid ancestor: the view taken will depend on whether the evolution of a coelom and metamerism is seen primarily as an aid to free-swimming locomotion or as an adaptation that improves burrowing. Nor is there agreement about the likely size of the annelid ancestor.

9.3 What are the advantages of the coelom and of metamerism?

9.3.1 The coelom

This cavity is useful in many ways. Separation of the gut and body wall enables each to move independently; the worm can bend without pushing food along the gut. A fluid-filled body cavity makes transport more effective, excretory organs, gonads and their products can be contained in the coelom. But probably the most important function of the coelom is to serve as a hydrostatic skeleton against which muscles can contract (see Chapter 5).

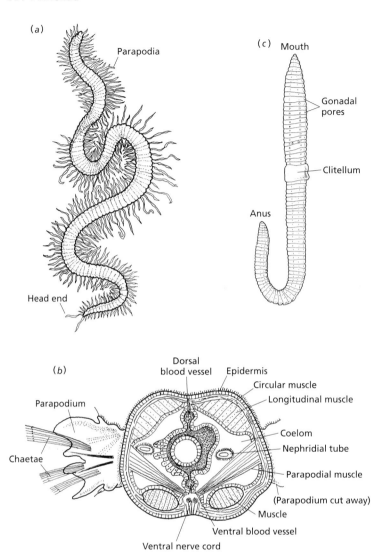

Figure 9.2. External views and transverse section (T.S.) diagrams to show the characteristic features of the three main classes of annelids: (a) *Trypanosyllis zebra*, a polychaete; (b) T.S. *Nereis*; (c) *Lumbricus terrestris,* an oligochaete;

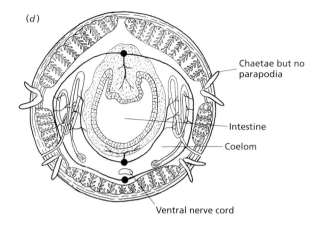

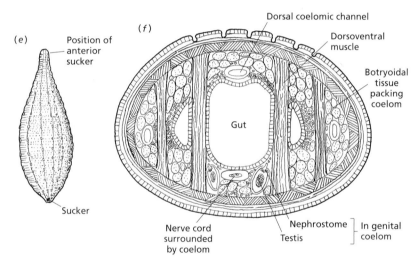

Figure 9.2 (cont.).   (*d*) T.S. *Lumbricus*; (*e*) a leech, viewed dorsally; (*f*) T.S. leech.

9.3.2 Metameric segmentation

Metameric segmentation also is above all concerned with movement. The primary metameric structures are the body wall muscles, innervated in annelids by segmental nerves arising from a chain of segmental nerve ganglia. Segmentation allows control of particular portions of muscle, enabling each to contract independently of its neighbour. When each muscle contracts in succession, a wave of contraction passes along the worm and can set up lateral undulations or peristaltic waves that can be used for burrowing. First a

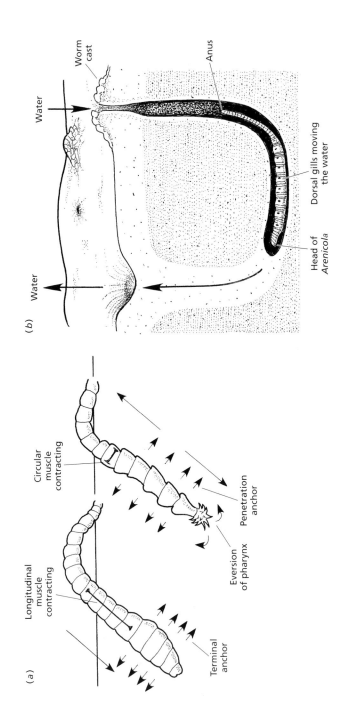

(a)

Longitudinal muscle contracting

Circular muscle contracting

Terminal anchor

Eversion of pharynx

Penetration anchor

(b)

Water

Water

Worm cast

Anus

Dorsal gills moving the water

Head of *Arenicola*

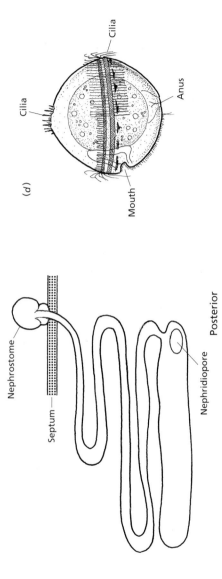

(c)

Anterior

Nephrostome

Septum

Nephridiopore

Posterior

(d)

Cilia

Cilia

Anus

Mouth

Figure 9.3. (a) The process of burrowing in *Arenicola marina*, with arrows indicating the direction of force; (b) *Arenicola* in its burrow, showing the water current; (c) the nephridium of an earthworm; (d) a polychaete trochophore larva.

'penetration anchor' makes and enters the burrow, and then a 'terminal anchor' takes hold and pulls in the rest of the worm (Figure 9.3*a*). Such a burrowing sequence is common in unsegmented animals also, but is very much faster and more efficient where fluid localised by septa in a front coelomic compartment can build up considerable hydrostatic pressure, as in earthworms. Burrowers in soft sand, such as lugworms, by contrast lack some septa and move coelomic fluid forward in bulk.

9.3.3 Locomotion

The slow walking of a typical polychaete such as *Nereis* is achieved by stepping with parapodia, their chaetae increasing traction with the substratum. Fast walking, however, uses the parapodia much less. The body is thrown into horizontal waves by the successive contraction and relaxation of the main body wall muscles alternately on the two sides; the lateral components of the force generated will cancel out and the worm moves forward by pressing backward on its environment (see Figure 5.1*a,b*). When the polychaete changes from fast walking to swimming, the very rapid wave illustrates the activity possible for a segmented coelomate worm.

Polychaete lateral undulation differs from wave propulsion in a nematode, eel or snake in that there are projecting parapodia. They do not contribute much directly, but their presence, spread out on the outside and clustered on the inside of each bend (see Figure 5.1*b*) means that the propulsive force is exerted on the outside convex surface rather than on the inside concave one. The wave of contraction has then to progress from the back to the front of the worm, unlike the front to back wave of a nematode, eel or snake, and the polychaete brain has to initiate locomotion at the far end of the body. A polychaete with all parapodia removed swims backwards.

Oligochaetes use the alternating contraction of circular and longitudinal muscles not to undulate but for peristalsis (see Figure 5.1*a*). Circular muscle contraction enables an earthworm to extend a longer thinner anterior portion while chaetae grip the ground in the portion just behind, made shorter and fatter by contraction of longitudinal muscles. The anterior circular muscles then relax and the wave of circular contraction passes backward along the worm. Leeches may 'loop' by attaching and detaching their suckers, or may swim by up-and-down waves propagated from the anterior end. To do this they need to flatten and stiffen their bodies by contraction of dorsoventral muscles, so that they can flex their muscles without the body shortening. The control of locomotion is more advanced than in other annelids, with a more condensed nervous system and more centralised functions.

9.4 How does a coelom introduce complexity?

The coelom both allows increase in size and activity and introduces a barrier between the outer body wall and the gut. A transport system therefore becomes essential, to supply the inner tissues with oxygen and the outer tissues with food.

9.4.1 The blood system

A small part of the primary body cavity or 'blastocoel' (see Box 5.3) of an annelid persists during development, becomes filled with blood and forms a closed system of vessels. General contractility of the dorsal longitudinal vessel pushes the blood forward, and then through a varying number of connecting vessels to flow posteriorly in the ventral vessel beneath the gut. There are further vessels, for example to the parapodia in polychaetes, and, while there is no true heart, there may be a number of blind-ending contractile vessels that assist blood flow.

Leeches have different circulatory systems. The coelom becomes packed with connective 'botryoidal' tissue leaving coelomic channels in which fluid circulates. The blood system then becomes much reduced or is absent.

9.4.2 Respiration

Permeability of the general outer surface is insufficient for most annelids to absorb enough oxygen, whether they are actively moving or living in a burrow or tube. Gills are required, such as parapodial lobes with a large surface area and internal blood supply. Tube-living polychaetes may need separate gills if the tentacles collect food from sand.

Respiratory pigments (see Box 9.2) are now necessary, in the blood to increase its oxygen carrying capacity or in tissues to facilitate diffusion. Haemoglobin is the commonest respiratory pigment, occurring in a very wide range of animals. It occurs in many annelids, but so also do two other rare iron-based pigments, haemerythrin and chlorocruorin, and the copper-based haemocyanin: no other phylum has so great a range.

Box 9.2
Respiration

Respiration is the oxidation of organic food with the release of energy. It is most commonly aerobic (using atmospheric oxygen) and has three stages:

1. **Gaseous exchange**: oxygen is taken in and carbon dioxide is given off at the respiratory surface, often assisted by ventilation movements.
2. **Transport of oxygen** from the respiratory organ to the cells, usually in the blood, and of carbon dioxide in the reverse direction (see Box 9.1).
3. **Within each cell** the biochemical processes are remarkably similar in all animals. Sugars are assembled and broken down by glycolysis to pyruvate, releasing energy that is incorporated into ATP or lost as heat. This is the **anaerobic** stage of respiration: in the absence of oxygen the pyruvate formed is converted to lactic acid. This incomplete breakdown may be the sole stage of cellular respiration in some animals or some circumstances, for example when muscles build up 'oxygen debt'. Most animals, however, oxidise pyruvate further (in the Krebs or tricarboxylic acid cycle) using molecular oxygen as the final hydrogen acceptor. This **aerobic respiration** makes much more energy available.

The overall equation is then

$$C_6H_{12}O_6 + 6O_2 = 6CO_2 + 6H_2O$$

Aquatic and aerial respiration

Oxygen may be obtained from water or air.

Advantages of respiring in water

1. The respiratory surface is supported by the water and will not dry out.
2. Carbon dioxide removal is very much easier: it is readily soluble, fresh water can hold about 3 volumes % CO_2 while air contains only 0.03 volumes %.

Advantages of respiring in air

1. The oxygen content of air is about 30 times that of water: air is 21% oxygen, while a litre of fresh water at 15°C will not contain more than 7.0 ml oxygen.
2. Oxygen diffuses 10000 times faster in air than in water.
3. Air, being less dense and viscous, moves faster than water over the respiratory surface.

Clearly, aquatic respiration requires a relatively large respiratory surface. In fresh water, however, there is an osmotic gradient across the surface (see Box 7.1), and increased permeability to oxygen (and therefore also to water) will increase the energy requirement for osmoregulation. Fresh water will in general be better oxygenated than the sea, because oxygen is less soluble in the presence of salts.

Oxygen tension

The availability of oxygen in solution is measured in pressure rather than volume. The 'partial pressure' of oxygen (that part of the total that is due to oxygen) is called the oxygen tension. Atmospheric pressure at sea level is measured in a barometer as 760 millimetres of mercury (mmHg) (in SI units, 101.3 kilopascals, kPa).

Air is about 21% oxygen, so that the oxygen tension in well-aerated sea water is approximately $0.21 \times 760, = 160\,\text{mmHg}$ (21.3 kPa).

Respiratory surfaces

The whole body surface

The surface of the body may be permeable to respiratory gases in marine animals, for example sponges, cnidarians, acoelomate worms, many small crustaceans and marine eggs and larvae.

Gills

The gills are thin-walled permeable extensions of the surface of an aquatic animal. They may be external, ventilated either by movements of water past the animal or of the animal through the water, or internal, when ventilation may require more energy. Gills and structures serving as gills include the parapodial lobes of free-swimming polychaetes and (where there are no separate gills) the

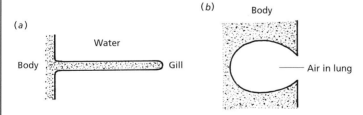

Diagram to show the basic difference between a gill and a lung.

Box 9.2 *(continued)*

tentacles of tube-living ones, the ctenidia of most molluscs and the mantle folds of limpets, all or parts of crustacean appendages and the tube feet of echinoderms.

Lungs

The lungs are internal air-filled sacs with an opening to the outside world and blood closely applied to their outer walls. In invertebrates, gaseous exchange is by diffusion without ventilation movements, for example in snails and slugs.

The tracheal systems

In insects and some other terrestrial arthropods, the tracheal systems consist of air-filled tubes branching all over the body and penetrating between the cells, providing extremely efficient aerial respiration for a small animal (see Chapter 15).

Respiration rate

The rate of respiration will vary according to the nature and circumstances of the invertebrate. It also will depend on the body size, activity and the ambient temperature. Respiration rate is usually equated with **metabolic rate**, the rate of the sum total of the activity of the animal. In aerobic animals metabolic rate can be measured by oxygen consumption per unit time; an alternative measure, applicable also to anaerobes, is the rate of heat production. Endothermic ('warm-blooded' vertebrates maintain a constant 'basal metabolic rate' at rest, but ectotherms (including all invertebrates) do not: they have the advantage of being able to 'shut down', economising on water and energy, in adverse conditions. In those invertebrates where (for a given temperature) a resting maintenance metabolic rate can be measured, it is greater in larger animals but not scaled up 1:1 with body mass. As in vertebrates, resting metabolic rate tends to be proportional to body mass to the power 0.75.

9.4.3 Excretion

While most polychaetes are osmoconformers, osmotic regulation is an important aspect of excretion in annelids able to colonise dilute environments; for example, *Nereis diversicolor* in estuaries and earthworms in wet soil can produce hypotonic urine (see Box 7.1). Nitrogenous waste may be excreted largely in the form of ammonia, or, where water is less freely available, specialised cells (in oligochaetes) or botryoidal cells (in leeches) may combine ammonia with carbon dioxide to make urea.

The excretory organs in larvae and some adult polychaetes (and possibly in ancestral annelids) are 'protonephridia', ectodermal tubes with blind ends (usually flame cells, see Box 9.2) where cilia-mediated filtration occurs. Some polychaetes have separate 'metanephridia', ectodermal ducts opening from the coelom by ciliated funnels. The tubes may lead separately to the outside but more often open into the mesodermal gonadal ducts occurring in most segments: the structure is then called a 'mixonephrium'. Oligochaetes and leeches have segmental separate metanephridia (Figure 9.3c), with gonads and their ducts restricted to a few anterior segments.

9.5 How do annelids reproduce and feed?

Polychaetes are by far the largest class of annelids, showing remarkably great diversity, while the oligochaetes (and still more the leeches) are more uniform. The division between polychaetes and oligochaetes is at first sight extremely tidy, with a habitat difference typically related to reproduction as follows:

9.5.1 Reproduction

Polychaetes are marine and have separate sexes, external sperm transfer and external fertilisation. The eggs are shed into the sea, where they develop into planktonic 'trochophore' larvae (Figure 9.3d). Oligochaetes, in fresh water or in soil, are cross-fertilised hermaphrodites with no external sperm transfer: the yolky eggs are fertilised within the worm or in the cocoon. The clitellum, situated near the female pores, secretes the cocoon into which (before or after fertilisation) the eggs are shed, and where they develop directly with no larval stage. Leeches in any habitat reproduce like oligochaetes, with direct development in a cocoon.

Generalisation provides a useful framework but (as so often in Biology) breaks down on account of the versatility of natural selection. Polychaetes are not all marine but have a few freshwater species, and rather more living under damp rotten logs on land; oligochaetes have a number of marine species. While the generalisations about reproduction and development in oligochaetes may hold, polychaetes vary widely. The female may collect and store sperm, for example, or may be viviparous; the trochophore larva may be modified or suppressed where development proceeds in a yolky egg. Asexual reproduction (with corresponding regenerative powers) may occur in polychaetes by budding, or may be combined with the sexual process as when the back end of a *Nereis* breaks off and becomes transformed into a pelagic 'heteronereis': like other pelagic polychaetes, it then floats due to its extended surface and has very large eyes. Gamete release may be synchronised in a swarm of sexually mature polychaetes.

9.5.2 Feeding and diversity

Polychaetes freely crawling or swimming in the sea are usually brightly coloured active predators, with biting jaws borne on an eversible pharynx. Most are carnivores but herbivores, detritivores and omnivores are known. There are very many genera, with names inspired by Greek nymphs and goddesses. *Nereis* is a familiar 'ragworm' as are *Nephtys* and *Phylloduce*; the Syllidae are slender with feathery parapodia (see Figure 9.2*a*), the Glyceridae have balloon-like pharynges, the Eunicidae have especially strong jaws and the scale worms such as *Aphrodite* have gill covers all down their backs. There are some 25 families, formerly grouped as 'Errantia', but they are not all closely related. The arrival of arthropod predators was no doubt a stimulus for rapid evolution among relatively defenceless worms; many times separately they took to living in burrows as sedimentary or detritus feeders.

External structures became reduced but muscles remained well developed and gills were needed. Examples include *Arenicola* the lugworm, responsible for worm casts on sandy and muddy beaches, and *Chaetopterus*, with remarkable regional differentiation of modified appendages that move water through its U-tube. Sessile tubicolous polychaetes, secreting the tube in which they live, are highly modified, with U-shaped guts. Well-developed nerves and muscles are important not for locomotion but for very fast retraction into the tube. The head becomes a crown of tentacles that may stretch over the sand and pick up particles, as in Terebellidae, or be held up, enabling ciliary feeding as in the fan worm *Sabella* and in the Serpulidae, makers of the calcareous tubes commonly found on rocks or seaweed (see Figure 9.4).

Diversity is slight among oligochaetes, which have a non-eversible pharynx and eat plant detritus, and even slighter among the leeches. Such is their morphological uniformity that Hirudinea all have 33 segments and their embryology is almost as rigidly determined by cell lineage as in nematodes.

9.6 How are annelids related to other phyla?

Annelid early development is unambiguously protostome (see Box 5.3). Cleavage is spiral, the blastopore becomes the mouth and the mesoderm is early set aside, later becoming blocks that form segments and split to make the coelom. Annelids are traditionally placed near the base of the protostome radiation and close to the molluscs; molecular methods support this placing.

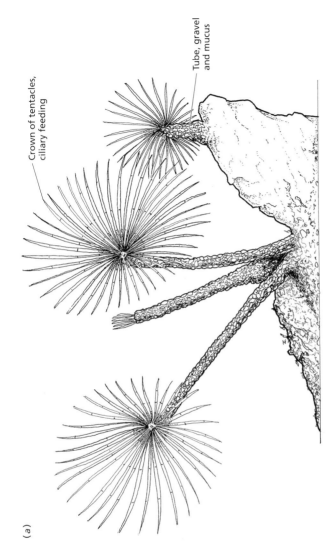

(a)

Crown of tentacles,
ciliary feeding

Tube, gravel
and mucus

Figure 9.4. Drawings of tubicolous polychaetes: (a) *Sabella*.

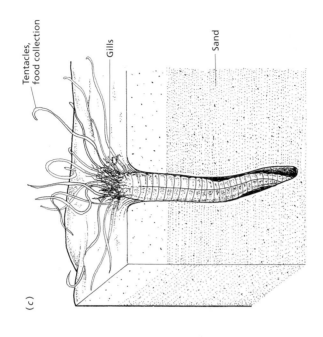

(c)

Tentacles, food collection

Gills

Sand

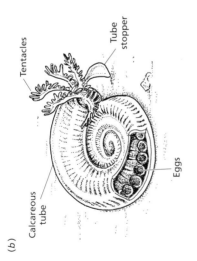

(b)

Tentacles

Tube stopper

Calcareous tube

Eggs

Figure 9.4 (cont.). (b) *Spirorbis*; (c) *Amphitrite*.

Box 9.3
Transport systems

The need for a transport system

Diffusion

The movement of a substance from a higher to a lower concentration of that substance is not a sufficient means of communication between the parts of any but the smallest animals. Planarians, for example, rely on diffusion for their oxygen supply and typically consume about 0.1 to 0.2 ml oxygen per gram per hour at 15 °C. Calculations based on the rate of diffusion suggests that planarians should be no more than 0.5 mm thick. [They do not all obey the rule. At this point biology takes over from physics: most planarians respire anaerobically part of the time, and much of the centre of the worm is food in the gut. The general point, however, is valid].

The transport or 'vascular' system

The commonest transport system is blood, a fluid tissue (i.e. it contains cells) either in closed vessels or in a haemocoelic cavity (see Box 5.2). Its function is to transport respiratory gases, food, excretory matter, hormones and many other substances. The assumption that the earliest blood systems evolved primarily for oxygen transport may be mistaken: nemertines, for example (see Chapter 7), have a blood system lying deep in the body and not mainly concerned with oxygen transport. Nemertines, unlike planarians, have a gut opening at both ends with one-way movement of food; this at once demands a transport system if food is to reach all parts of the body. Even in planarians where the mouth is the only opening, food would not reach all tissues but for the considerable branching of the gut.

Transport of oxygen and carbon dioxide

Respiratory pigments

Pigments that combine reversibly with oxygen are in most animals a necessary aid to respiration. By far the commonest is **haemoglobin** (Hb) found at least in some members of almost every phylum. Structurally a protein containing iron and haem, it is related to the cytochrome respiratory enzymes used in cellular respiration in all animals. Haemoglobin may occur in solution or in corpuscles, where it can become very much more concentrated without raising the osmotic pressure of the blood. Primarily it transports oxygen, but it may be an oxygen store or, often as a simpler form of the molecule called 'myoglobin', it may facilitate oxygen diffusion.

Box 9.3 *(continued)*

Haemocyanin is a copper-based pigment occurring in many arthropods and molluscs. It has a lower oxygen-carrying capacity than haemoglobin and cannot be contained in corpuscles. Other respiratory pigments such as **haemerythrin** and **chlorocruorin** occur, especially in annelids.

Carbon dioxide

Carbon dioxide is also transported in reversible combination with haemoglobin, as bicarbonate (HCO_3^-) ions or in physical solution.

Oxygen dissociation curves

Graphs showing oxygen dissociation (see diagram) express the relationship between the amount of oxygen available and the amount taken up by the blood. They are informative about the haemoglobins in different animals or in different circumstances. The characteristic sigmoid curve may be shifted either to the left, when oxygen can more easily be taken up, or to the right, when oxygen is more easily given off in the tissues. The 'Bohr Shift' occurs in most animals: in diagram (*a*) curve A represents conditions at the respiratory surface and the shift to the right (to curve B) is caused by the higher carbon dioxide concentration (greater acidity) at the tissues, enabling more oxygen to be given off from the haemoglobin molecule. Animals living in places with low oxygen tensions, where ease of oxygen uptake is paramount, have curves well to the left (see diagram (*b*) curve C, for the polychaete *Arenicola* in its burrow). Other factors affecting the balance of advantage between ease of uptake and ease of release of oxygen are temperature (see curves D and E for the earthworm *Lumbricus*) and high altitude, where there is less total oxygen due to low atmospheric pressure (and the curve moves to the left).

Countercurrent systems

Where the direction of flow is opposite in closely apposed channels, countercurrent systems increase the uptake of oxygen at respiratory surfaces. With counter-current flow there is a concentration gradient of oxygen (or any other substance, or heat) along the whole surface in contact, allowing very much more diffusion than can occur in parallel ('co-current') flow where there is a high initial gradient progressively decreasing (see diagrams (*c*) and (*d*)).

The transport systems of different invertebrates

Absence of transport systems

There is no transport system in sponges or in most cnidarians, where every cell is close to the ambient medium. None is present either in platyhelminths, gnathostomulans, gastrotrichs, nematodes (where fluid under high pressure is

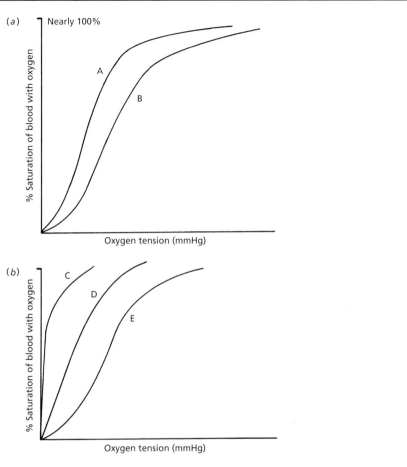

(a) and (b) Oxygen dissociation curves: at (A) the respiratory surface and (B) the internal tissues; for (C) *Arenicola* within its burrow, (D) *Lumbricus* at 7°C, (E) *Lumbricus* at 20°C.

moved in the pseudocoel) nematomorphs, rotifers, acanthocephalans, kinorhynchs, priapulids (though there are cells containing haemerythrin in the body cavity), entoprocts, sipunculans, bryozoans or chaetognaths.

Sea water transport systems

Sea water channels occur in large jellyfish (Scyphozoa). Although both cell layers are in contact with the sea for respiratory exchange, very thick mesoglea may separate the mouth from the ectodermal structures. Food in sea water circulates in canals through the jelly: straight radial ones from mouth to margin and much branching canals where water returns inward, as can often be seen through the translucent jelly (see Figure 4.5).

Box 9.3 (continued)

(c)

(d)

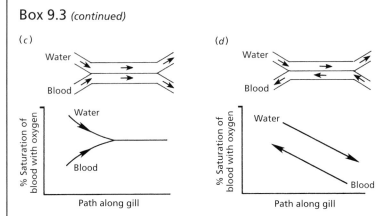

(c) Co-current and (d) countercurrent flow of water and blood at a gill, showing gradients of oxygen content as the water passes along the gill.

Closed blood systems

Systems where blood circulates (forward dorsally and backward ventrally, except in chordates where it is the reverse) in narrow vessels, propelled by hearts or general contractility of the main vessel, occur in annelids (see Figure 9.2), echiurans, pogonophorans, phorona and chordates; nemertines have a differently derived closed system, and in the very active cephalopod molluscs the open haemocoel has become canalised into vessels, enabling faster circulation.

Circulation in channels of the coelom

Coelomic circulation occurs in leeches, where the coelom is invaded by connective tissue or 'botryoidal' tissue, leaving a system of sinuses and channels where fluid circulates, propelled by contractions of the main lateral vessels. The blood system is much reduced or absent.

Haemocoels

Open blood systems occur in molluscs, brachiopods and arthropods (see Figure 12.4). Insects have a tracheal system that brings oxygen to every cell. The blood is not involved in respiration and (except in a few larvae) there are no respiratory pigments.

10

Mollusca: general and Gastropoda

'Life in a shell' is the molluscan theme; it sounds very restricted. Yet Mollusca constitute a remarkably large and wide-ranging phylum of at least 100000 species, mostly marine but including many fresh water species and about 10000 land snails. The shell is used in a variety of ways, or discarded altogether. The molluscan body plan is unique, and can be traced in animals as different as snails, scallops and squids, functioning over an unusually large size range (from a few millimetres to 10 metres). This account first introduces the distinctive body plan and explains how the shell may be used and then, for each of the main groups, indicates how the basic design has become modified and how so many different molluscs manage to make a living.

10.1 What is the basic molluscan body plan?

As is shown in Figure 10.1*a*, the internal organs ('viscera') are contained between a ventral muscular creeping foot and a dorsal calcareous shell secreted by an underlying epidermal covering, the mantle. The mouth opens anteriorly into a buccal cavity with a small sac containing the radula, a rasping tongue covered in teeth made of chitin, a distinctive feature unique to molluscs. The mantle tends to overhang the body as a pair of double-walled folds, leaving posteriorly a space filled with ambient water. This mantle cavity contains the 'ctenidium' (morphological term) or 'gill' (a name indicating its respiratory function). The gut bears a digestive gland, often the site both of enzyme secretion and food absorption: the anus opens into the mantle cavity. The body is unsegmented and the main body cavity is a haemocoel (see Box 5.3). There is also a very much restricted coelom, consisting of a space round

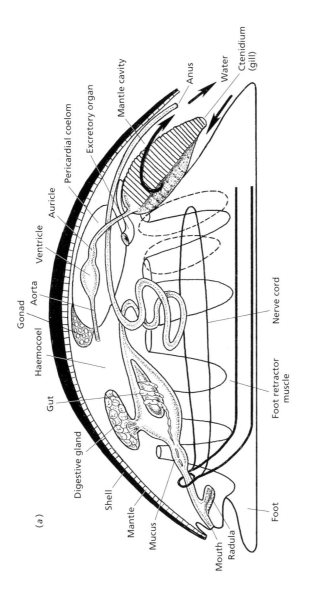

(a)

Gonad

Haemocoel

Gut

Digestive gland

Shell

Mantle

Mucus

Mouth

Radula

Foot

Nerve cord

Foot retractor
muscle

Aorta

Ventricle

Auricle

Pericardial coelom

Excretory organ

Mantle cavity

Anus

Water

Ctenidium
(gill)

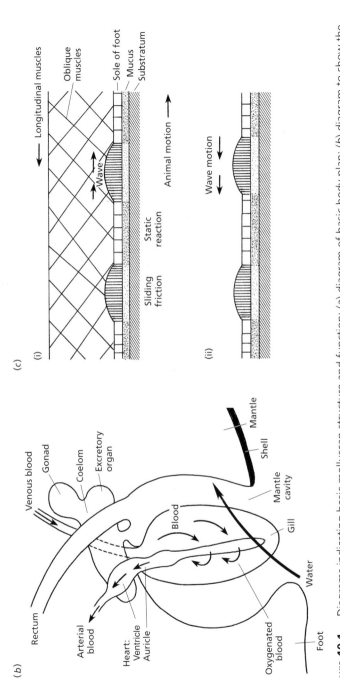

Figure 10.1. Diagrams indicating basic molluscan structure and function: (*a*) diagram of basic body plan; (*b*) diagram to show the water flow over the gill, the blood circulation and the relationship of the heart and excretory organs, in a primitive gastropod; (*c*) the locomotor muscular waves over the molluscan foot, (i) direct waves, in the direction of motion, and (ii) retrograde waves, in the opposite direction to the motion.

(*b*)

Venous blood
Gonad
Coelom
Excretory organ
Rectum
Blood
Arterial blood
Heart:
Ventricle
Auricle
Oxygenated blood
Foot
Water
Gill
Mantle cavity
Shell
Mantle

(*c*)

(i)

Longitudinal muscles
Oblique muscles
Sole of foot
Mucus
Substratum
Wave
Animal motion
Sliding friction
Static reaction

(ii)

Wave motion

the heart and the cavities of the excretory and reproductive organs. The brain primitively is a simple ring round the oesophagus, with two longitudinal nerve cords. Sensory tentacles are usually present and there may be eyes. Development is by spiral cleavage and coelom formation is schizocoelic, i.e. molluscs are typical protostomes (see Box 5.3). There may be a trochophore-type larva (as in polychaetes) followed by a 'veliger' with a shell rudiment.

10.2 How can such an animal function?

10.2.1 Feeding: the radula

The radula is unique to molluscs, and occurs in all groups except bivalves, which have abandoned feeding with a rasping tongue. The radula consists of chitinous teeth stretched over a supporting belt, pointing backwards. The teeth are in horizontal rows, shapes varying according to their position, and are continuously worn away and replaced as the supporting belt moves forward. The radula is primitively a scraping organ used when collecting algae and other small food organisms from hard surfaces, but has become adapted for a wide range of diets (see below).

10.2.2 Respiration: the ctenidium (gill)

Typically a pair of ctenidia hang in the ciliated mantle cavity. Each ctenidium has a central axis with a lamella on each side, and is covered in cilia. Water carrying oxygen is driven by ciliary action to enter the mantle cavity ventrally, make a U-turn over the gill and exit dorsoposteriorly: it therefore flows over the gill from front to back (Figure 10.1*b*). The blood inside the gill flows across in the opposite direction as shown. The blood, with its affinity for oxygen increased by haemocyanin in solution, receives oxygen by diffusion from the water stream and is collected into a central vessel. Sensory organs beside the gill monitor the quality of the incoming water; the exhalant water stream removes waste and any genital products from the mantle cavity.

10.2.3 Blood circulation: the haemocoel

Oxygenated (arterial) blood from the ctenidium is pulled into the contracting heart (Figure 10.1*b*), and from the ventricle is discharged into the open space of the haemocoel, where it bathes the tissues. Circulation is slow. Deoxygenated (venous) blood enters the base of the ctenidium, becomes oxygenated as it passes through the small vessels in the lamellae, then continues by way of the larger central vessel into the ventricle.

10.2.4 Excretion: the coelom

The excretory organs (kidneys) of molluscs are mesodermal coelomoducts (see Box 9.1). Waste collected from the haemocoel filters into the pericardial coelom: this is the largest part of the molluscan coelom, between the heart and its enclosing organ, the pericardium (see Figure 10.1b). Waste is then discharged into the mantle cavity through a duct which may also carry the gametes.

10.2.5 Locomotion: the foot

The foot is strongly muscular. The primitive ventral creeping foot moves the animal forward by waves of muscle contraction working against fluid pressure from neighbouring cells and from the haemocoelic spaces within the foot. Production of mucus is a necessary aid to locomotion. The muscle waves may either be 'direct' (in the direction of locomotion, as in polychaetes) or 'retrograde' (in the opposite direction, as in nematodes) according to the stage in the contraction cycle at which the foot is stationary on the ground (Figure 10.1c). Very small molluscs may be propelled by cilia on the surface of the foot rather than by muscular waves.

10.3 What is the shell and how may it be used?

The shell consists of two or more layers of calcium carbonate deposited within a framework of a protein called conchiolin, and is covered externally by a different protein layer called the ' periostracum'. The calcium carbonate is derived from calcium bicarbonate in solution and laid down as crystals of calcite or aragonite. The shell is secreted by the mantle, mainly at the periphery where the outermost layers are deposited, but there may be some thickening over the general mantle surface. Shells are stiffer and weaker than our bones: stiffer because of the high proportion of inorganic crystals and weaker because there is less of any strengthening protein. The innermost 'nacreous' (mother o' pearl) layer has sheets of aragonite all orientated parallel to the surface, and is stronger than the outer 'prismatic' layer made of crystals. The horny outer periostracum protects the calcium carbonate from carbonic acid and other agents of damage.

Unlike the arthropod exoskeleton, the molluscan shell does not fit the body very closely and is not moulted as the animal grows. It may support the body but does not provide its framework. It is not jointed and does not primarily serve to attach muscles. Primitively the shell was probably a shield protecting the soft-bodied mollusc against predators, mechanical damage and (out of the sea) against desiccation. It may be modified into protective housing that can completely enclose or be pulled down over the soft body, or the body can be pulled into the shell, by contraction of pedal retractor muscles originating on the inner shell surface and attached to the foot. The prevalence of molluscs in the intertidal region and of snails on land suggests that the protection afforded by the shell has been very important.

The shell has many other functions in different molluscs. Muscles may be attached to it, for example those effecting torsion in gastropods (see below) and the adductors holding together the two halves of bivalves. It can be used for burrowing or boring. It may form a keel in pelagic molluscs or it may contain gas and assist buoyancy. Yet all groups except bivalves have a tendency for evolutionary loss of the shell. Loss of protection is balanced by the gain in mobility and lightness, and in different circumstances either advantage may be paramount. For example, snails are well protected above ground. Slugs can obtain comparable protection from desiccation and predators only by burrowing beneath the surface, as their shape and lack of shell allows. Snails are strikingly plentiful and diverse on calcareous soils: where there is less calcium available, slugs may be commoner.

10.4 What are the main groups of molluscs?

See Figures 10.2 to 10.5 for those discussed in this Chapter.
- **Aculifera** with spicules.

  - **Aplacophora**: worm-like molluscs, without shells.

    - **Chaetodermomorpha** with no foot; burrowing.
    - **Neomeniomorpha** with ridge representing foot; creeping.

  - **Polyplacophora**: Chitons, with eight dorsal shell plates and spicules in the surrounding mantle. Ventral creeping foot.

— **Conchifera** with shells, not spicules. Show typical molluscan body plan.

  — **Monoplacophora**: shell a single cap or cone.
  — **Gastropoda**: shell single, often coiled. Torsion during development brings mantle cavity to the front.

    — **Prosobranchiata**: gill in front. A very large diverse group including limpets, winkles, whelks: mostly marine.
    — **Opisthobranchiata**: gill behind, due to detorsion associated with loss of shell. Includes sea-slugs.
    — **Pulmonata**: terrestrial. The gill is lost and the mantle cavity becomes an air-filled 'lung'. Land snails and slugs.

  — **Bivalvia**: shell as two lateral valves, hinged at the top, enclosing the body. Typically sedentary. Clams, mussels, oysters, scallops.
  — **Scaphopoda**: shell tubular, opening at both ends: a small and uniform marine Class.
  — **Cephalopoda**: actively carnivorous marine molluscs.
    — **Ammonoidea** with coiled external shell. Extinct ammonites, common as fossils.
    — **Nautiloidea** with coiled external shell, including present-day forms.
    — **Coleoidea**: much modified molluscs with highly developed brains, sense organs and nervous systems, swimming fast by jet propulsion. The shell is either internal and reduced or lost entirely. Cuttlefish, squids, octopus.

10.5 What are the evolutionary relationships of molluscs?

10.5.1 Relationship with other phyla

Traditionally, the resemblance of molluscs to annelids in early development and their possession of a polychaete-like trochophore (Figure 10.2*a*) has been seen as overwhelming evidence for a close relationship between these two protostome phyla, despite longstanding arguments about the equivalence of the coelom in annelids and molluscs and whether molluscs were primitively segmented. More recently, emphasis has shifted to the adult forms and it has become commoner to derive molluscs directly from ancestral flatworms, creeping forms that are assumed to have evolved a protective shell dorsally and then to have needed a more complex structure. However,

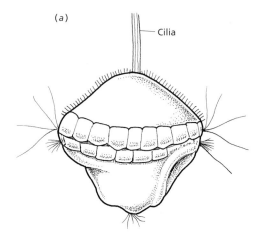

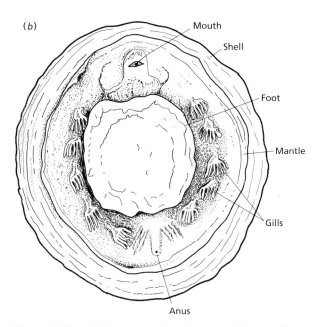

Figure 10.2. (a) A gastropod trochophore; (b) *Neopilina*, a monoplacophoran, in ventral view;

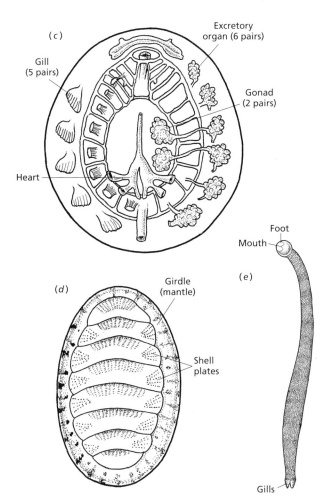

Figure 10.2 (*cont.*).   (c) diagram of *Neopilina* showing the multiplication of organs; (d) *Chaetopleura*, a chiton (polyplacophoran); (e) *Chaetoderma,* an aplacophoran.

comparison with present-day flatworms is not compelling, and molecular evidence places molluscs much nearer to annelids than to platyhelminths (see Chapter 20).

10.5.2 Relationships within the phylum

Until it is decided whether the primitive mollusc was a coelomate, segmented perhaps and burrowing, or an acoelomate, non-segmented gliding flatworm-with-a-shell, it is difficult to root the phylum as a whole. Monoplacophora, Polyplacophora and Aplacophora are all too specialised to be convincing candidates for the primitive position.

10.5.3 Monoplacophora

These molluscs with the shell as a single cap or cone, were known only as fossils until a deep sea dredge in 1954 brought up *Neopilina* (Figure 10.2*b*). Although simple in structure it appeared to have two extensive dorsal coeloms and some structures were multiplied (five pairs of gills, six pairs of excretory organs, eight pairs of pedal retractors and two pairs of gonads). *Neopilina* was hailed as close to the ancestral mollusc, with a well-developed coelom and, although the numbers did not coincide, vestigial segmentation. Since then ten more species have been found, none longer than 30 mm, living as specialist detritus feeders in the ooze at the bottom of very deep seas. Further study revealed that the 'coelomic' cavities were linked to the pharynx, and the (varying) multiplication of some organs is now seen as a specialised feature. The Monoplacophora are concluded to be an evolutionary dead-end and not to be close to the base of the molluscs.

10.5.4 Aplacophora

These worm-like forms with spicules but no shell (Figure 10.2*e*) are widespread in the deep sea and also occur in shallower water. They are up to 300 mm in length; some 288 species are known. They possess a radula, which is diagnostic of molluscs. Rather than representatives of the primitive condition, they are now thought to be secondarily worm-like, perhaps on account of early reproduction while still in the larval form.

10.5.5 Polyplacophora

The Polyplacophora were once grouped with Aplacophora as 'Amphineura', since both groups, and also Monoplacophora, have a ladder-like construction of the paired nerve cords. Polyplacophora are now seen as separate. The eight overlapping dorsal shell plates, surrounded by a spicule-bearing 'girdle' or mantle (Figure 10.2*d*) are quite unlike any other molluscan shell: each plate is penetrated by living tissue with sensory structures and nerves, and has a separate pedal retractor muscle. Their embryological formation also is different: the plates are laid down as rods in dorsal horizontal grooves.

Polyplacophora are known from the early Cambrian (in Australia) and there are about 800 species of living chitons, over half living in shallow waters. They are common in the intertidal region (in the southern hemisphere the shores are strikingly rich in large chitons). They creep over rocks, grazing mainly on algae. Small gills are multiplied in rows within the mantle cavity, which extends on either side of the powerful foot, which by contraction of the pedal retractors can clamp the animal down on to the rock.

In conclusion, Polyplacophora constitute a successful group of molluscs but, like the Monoplacophora and the Aplacophora, are in evolutionary isolation. It is hard to see how the shell plates could be derived from, or give rise to, the shell of other groups. For Molluscs as a whole, the verdict must be that the seven Classes are distinct and clearly related to each other but we do not yet know how.

Gastropods, bivalves and cephalopods are the dominating Classes with which this account is primarily concerned.

GASTROPODA

Gastropods are the largest and most diverse group of molluscs (some 50 000 species have been described). They occur at all levels of the sea, in fresh water and on land, with a remarkable range of feeding methods, even including internal parasitism. The impression is 'You name it, some gastropod does it'. This account cannot do justice to such diversity; it can only indicate its basis.

10.6 How is the molluscan body plan modified in gastropods?

The head is usually well developed with eyes and sensory tentacles. The primitive ventral creeping foot is usually retained, and the 'visceral hump' includes the digestive gland and part of the gut. The shell is in one piece, enclosing the visceral hump and the mantle cavity, all as represented in Figures 10.1*a* and 10.3.

10.6.1 Shell coiling

The shell is often coiled. Development as an expanding cone would make the visceral hump large and unwieldy unless it were coiled, and the shell becomes coiled also.

10.6.2 Torsion

The outstanding characteristic of gastropods, however, is torsion, a phenomenon entirely separate from coiling of the visceral hump and of the shell. At the veliger larval stage the visceral hump is twisted through 180°, bringing the mantle cavity to the front (hence the name given to the basal group of gastropods, 'Prosobranchs', see Figure 10.4). The foot (except in pulmonates) carries an operculum, which can close the opening when the body is withdrawn into the shell.

Torsion through 90° is achieved in a few hours, by contraction of an asymmetrical muscle attached to the right side of the shell and the left side of the head and foot (Figure 10.4*a,b*). At the point of twisting the two main nerve cords are crossed, making a figure of 8. Torsion is completed more slowly, by asymmetrical growth of the larva. What is the adaptive significance of this curious phenomenon? Possibly it helps the mollusc when attacked to draw the head into the shell before the back end of the foot, but that is not very convincing, a small larva is likely to be swallowed in one gulp. A more satisfactory theory concerns the water currents in the mantle cavity. Where the mantle cavity is posterior and overhung by the shell, it may be difficult to achieve a respiratory current flowing in the opposite direction to the current flowing past a forward-moving animal (Figure 10.4*c*). After torsion this difficulty disappears, as fresh clean water flows readily into an anterior mantle cavity. A new problem, however, arises: with the anus now anterior, waste will be deposited on the gill. This problem is reduced if the shell develops a hole or a series of holes: the anus can be retracted and the outflowing respiratory current will carry the faeces (Figure 10.4*d*). This condition obtains in the more primitive prosobranchs, for example the abalone (ormer) *Haliotis*, the slit shell *Emarginula* and the keyhole limpet *Diodora*. True limpets that clamp tightly

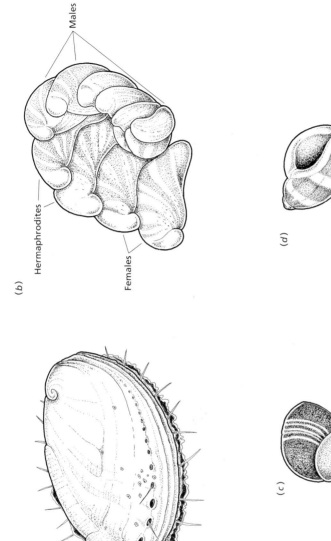

Figure 10.3. Drawings of some prosobranch gastropods: (*a*) *Haliotis*, an ormer; (*b*) *Crepidula*, the slipper limpet, showing the gradation of the sexes; (*c*) *Littorina*, a periwinkle; (*d*) *Nucella*, the dog whelk.

(*a*)

(*b*)

Males

Hermaphrodites

Females

(*c*)

(*d*)

Siphonal groove

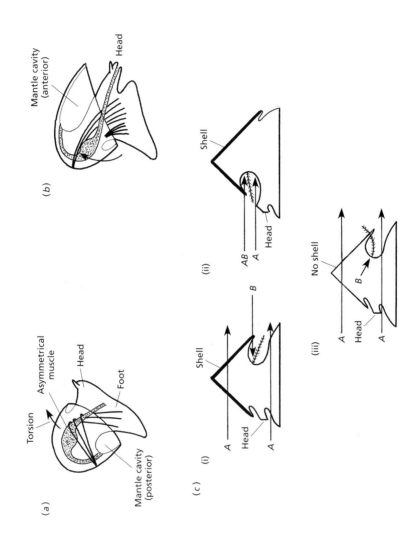

(a)

Torsion

Asymmetrical muscle

Head

Foot

Mantle cavity (posterior)

(b)

Mantle cavity (anterior)

Head

(c)

(i)

Shell

A

Head

A

B

(ii)

Shell

AB

A

Head

(iii)

A

Head

A

No shell

B

on the rocks, for example *Patella*, do not have this problem, since the
cavity is greatly reduced and the ctenidia are entirely lost: respiratory
ge occurs in folds of the mantle edge ('pallial gills'). The vast majority
sobranchs (including periwinkles, cowries and conch shells) have lost
enidium and draw in water from the side, first over the single gill then
er the anus (Figure 10.4*e*). The most advanced prosobranchs are the car-
us whelks and dog whelks, also with a single ctenidium. They are rec-
ble by a groove at the shell opening, where a siphon is projected as they
or prey.

3 Opisthobranchs

nembers of this group have untwisted: the mantle cavity is again at
ck. That this process is truly a secondary 'detorsion' is shown by the
ny of the nerve cords, remnants of torsion in some species and the
on of a single ctenidium. Detorsion has occurred many times in par-
n combination with the loss of the shell (without a shell, clean water
reaches a posterior mantle cavity, see Figure 10.4*c*). Opisthobranchs
e the sea hare *Aplysia* (Figure 10.5*a*), much used in neurobiology on
nt of the simple accessible nervous system and the large nerve cells, the
tterflies, with or without shells, that swim by means of a modified foot
e sea slugs, for example *Archidoris* and *Aeolidia* (Figure 10.5*b,c*).

4 Pulmonates

s group the mantle cavity has become an air-filled lung (Figure 10.5*d*)
ey may also show anatomical evidence of detorsion. The many fresh-
pulmonates either carry an air store, frequently replenished at the
e, or, as in the freshwater snail *Planorbis*, they have secondarily developed

How may gastropods feed?

tively, small particles of food are scraped off hard surfaces by the radula,
curs in browsers and grazers such as limpets, top shells and periwinkles
nany other intertidal prosobranchs. Small food particles are also collected
ny other ways, mainly in the sea: detritus feeders extract decomposing
ic material from the sea bottom; suspension feeders include those trap-

down
mant
excha
of p
one
out c
nivo
ogni
hum

10

The
the
ana
ret
all
rea
in
ac
se
ar

1

l
a
v
s
9

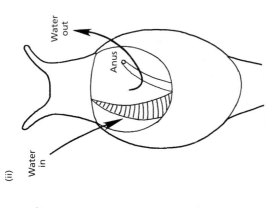

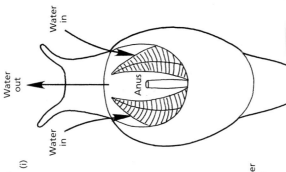

(e) (i)

(ii)

(d)

ping food in mucus such as the pelagic sea butterflies and ciliary feeders such as *Crepidula*, which uses enlarged ctenidia with many ciliary tracts (much as in bivalves).

Many gastropods, however, use the radula to obtain larger pieces of food, as herbivores, scavengers or carnivores. The radula may cut, pierce, grasp or even masticate the food and there may in addition be jaws. Herbivores include most pulmonates, opisthobranchs such as *Aplysia* and many freshwater prosobranchs. Carnivores include the cowries browsing on ascidians, and neogastropods that use a proboscis to find and seize their prey. Of these, whelks may use the foot to smother the victim or to wedge open the valves of a bivalve: *Urosalpinx* in oyster beds secretes acid, which assists penetration of the oyster shell by a drill-like radula: cone shells have poisonous radula teeth and a bite may be fatal to humans. The aeolid opisthobranchs (see Figure 10.5c) live and feed on hydroids, anemones or even corals, and may resemble them so closely that they are very hard to see. The cnidarian stinging cells are not digested but are passed from the gut to 'tentacles' on the sea-slug's back, where they protect the shell-less animal. Pulmonates may eat earthworms or slugs: the slug *Testicella* seizes earthworms, grips them with the radula and swallows them whole.

Ectoparasites include *Thyca*, a prosobranch that has lost the radula, with the mouth modified into a sucker by which it attaches itself to a starfish and extracts soft tissue. Endoparasites occur in one prosobranch family, for example the worm-like *Enteroxenus* without a shell that lives in sea cucumbers and absorbs food all over its surface.

10.7.1 Digestion

Food is moved along the gut by the beating of cilia rather than by muscular contraction. Usually the stomach is the site of extracellular digestion, by enzymes from the salivary glands. The food is then sorted in the caecum: the fine particles pass to the main digestive gland, where there may be intracellular digestion as well as absorption. Heavy rejected particles pass to the intestine, being stored there when the animal is withdrawn into the shell, or voided at the anus.

10.8 Why are many gastropods hermaphrodites?

Adaptations for sexual reproduction in gastropods illustrate the balance of advantage between having separate sexes or having the organs of sexual

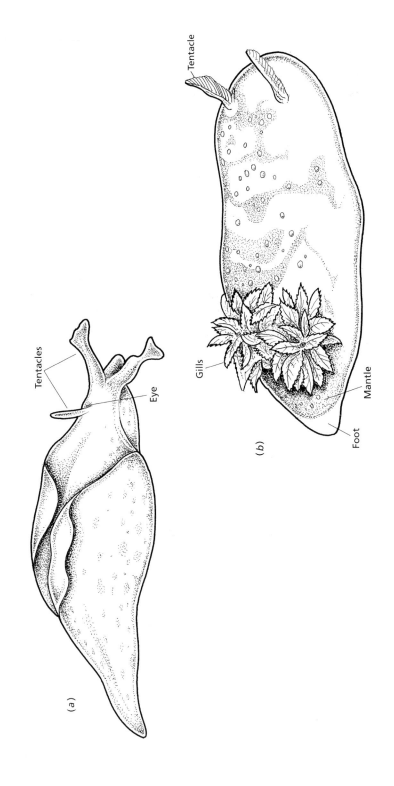

Tentacle

Tentacles

Eye

Gills

Mantle

Foot

(a)

(b)

Figure 10.5. Drawings of some opisthobranchs and a pulmonate: (*a*) *Aplysia*, the sea hare; (*b*) *Archidoris*, an opisthobranch; (*c*) *Aeolidia papillosa*, an aeolid opisthobranch; (*d*) *Helicella*, a pulmonate snail.

Tentacles

Eye

Lobe with
nematocysts

(c)

(d)

reproduction in one body (i.e. being hermaphroditic). Animals with separate sexes have the economy of producing only one set of sexual organs and the opportunity for advantageous division of labour between a mobile male and a food-storing female. To have separate sexes is by far the commoner reproductive strategy, especially in the sea. Hermaphrodites may, however, have the advantage where the opportunities for animals to meet is reduced, since every meeting is a potential mating. Cross-fertilisation is advantageous in that it introduces genetic variation, but if no meetings occur, self-fertilisation may be possible for a hermaphrodite. Sessile animals are usually hermaphroditic. Colonisation of land or even fresh water, especially in isolated or newly colonised habitats, may separate individuals, in particular if they are slow-moving, like molluscs. Further, sense organs evolved in the sea may not work so well for finding mates on land.

Most marine prosobranchs have separate sexes. Primitively the gonads have no separate ducts, but shed their products into the pericardial coelom or directly to one of the excretory ducts, which carry them to the mantle cavity. Eggs and sperm are washed out to sea and fertilisation is external. More advanced prosobranchs have gonadal ducts, and a penis in the male, allowing internal fertilisation. Some species are parthenogenetic: they have no males and the eggs develop without being fertilised, for example the common small freshwater prosobranch *Potamopyrgus jenkinsi*, an immigrant from New Zealand. A few prosobranchs are hermaphroditic (including the sessile limpets) and most hermaphrodites are protandrous (i.e. the male organs develop first. A striking example is the slipper limpet. *Crepidula fornicata* (see Figure 10.3*b*), where individuals live stacked up in a pile ('fornicata' means 'arch-forming'). All the young are at first solitary and male: they become hermaphroditic and then female. Other young males arrive and make a pile. They begin to feminise, but the first female produces a hormone (called a 'pheromone', since it acts outside her body) that maintains the masculinity of any male in her vicinity. In a typical pile the bottom (oldest) animals will be female, the next one or two hermaphrodite and the youngest male. All the females can then be cross-fertilised.

Opisthobranchs and pulmonates are all hermaphrodites. Opisthobranchs are slow-moving animals with specialised feeding habitats, giving them a patchy distribution. The land-living pulmonates are cross-fertilised hermaphrodites. There is an ovotestis discharging into a common hermaphroditic duct, and separate sperm ducts and oviducts. Various adaptations promote copulation, and self-fertilisation occurs in a very few species. As in all freshwater gastropods and most terrestrial invertebrates, the free-living larva has been suppressed and there is direct development within large yolky eggs.

10.9 Conclusion

Gastropods show greater diversity than any other non – arthropod group of invertebrates. From a relatively unspecialised starting point, adaptive radiation (i.e. evolutionary divergence) has been most pronounced. At the same time, gastropods also provide very many clear examples of evolutionary convergence: in all groups similar changes have occurred from many different starting points; for example, the changes associated with torsion in prosobranchs, the loss of shell combined with detorsion in opisthobranchs (in six different groups independently) and changes in shell shape in land snails. The molluscan plan has already been shown to be most successful.

The next chapter considers the other two main groups of molluscs, which are very different from each other – Bivalvia and Cephalopoda – illustrating extremes of divergence in the evolution of the molluscan body plan.

11

Mollusca: Bivalvia and Cephalopoda

11.1 How is the molluscan body plan modified in bivalves?

The molluscan body is laterally compressed between the two halves of the dorsally-hinged shell. Characteristically the ctenidia are greatly enlarged, with elaborate ciliary tracts used for filter-feeding. The head and its associated sensory organs have been lost, as also has the radula, and feeding is assisted by ciliated 'labial palps'. The foot is lifted off the ground, becoming wedge shaped: in sedentary forms, glands in the foot secrete a 'byssus' of threads attaching the bivalve to the substratum (Figure 11.1*a*). Muscles are strongly developed: a pair of adductors attached to the inner surfaces of the shell hold the two halves together, working against the elastic recoil of the hinge ligament, and, if there is a byssus, anterior and posterior retractor muscles hold it in place (Figure 11.1*b*). Inhalant and exhalant siphons determine the flow of water in the mantle cavity (Figure 11.1*c,d*).

11.2 What is the range of bivalves?

They are very common animals, with about 8000 known species. All are aquatic and most are marine, perhaps because the very large permeable gill surface presents a considerable osmotic problem for a freshwater bivalve. The common mussel *Mytilus* extends into estuaries, but few species live in fully fresh water. In the absence of a powerful kidney, a common adaptation is dilution of the internal body fluid and thus reduction of the osmotic gradient

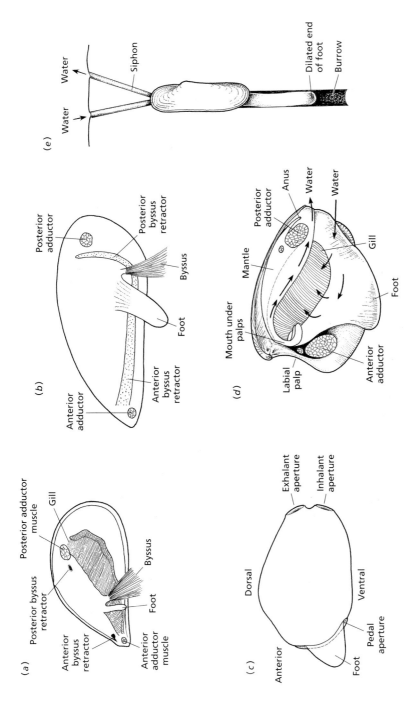

Figure 11.1. The structure of bivalves: (*a*) lateral view of *Mytilus* with one shell valve removed; (*b*) a bivalve with parts removed, to show the main muscles; (*c*) a bivalve in its mantle, with both shell valves removed; (*d*) lateral view of *Mercenaria* to show water currents in the mantle cavity; (*e*) a razor shell burrowing, to show the long siphons.

across the gills. The most dilute animal known is the freshwater clam *Anodonta*, with an internal fluid concentration one-tenth of that of a freshwater crayfish.

In size, bivalves range from a few millimetres to the giant clam *Tridacna*, which is over a metre long. Here the exposed mantle edge contains symbiotic algae. Like those of the corals among which the clam lives, the algae may assist precipitation of calcium carbonate (needed to make the large shell) by removing carbon dioxide and reducing acidity.

The activity range is considerable also. Most bivalves are sedentary, i.e. buried or lying on a substrate but capable of limited movement. Some are sessile, i.e. entirely incapable of locomotion, such as the oysters, which are cemented to hard surfaces by one shell valve. Young stages are motile, i.e. capable of moving from place to place: planktonic larvae are important agents of dispersal, and even after metamorphosis young bivalves may move by extending the foot, attaching the cup-like sucker at its tip to the substratum and pulling the rest of the body over the foot. A similar technique is used by burrowing bivalves, such as razor shells (which burrow remarkably rapidly) and clams: the foot penetrates the sand, swells at the tip and takes hold, then the longitudinal muscles of the foot contract. In burrowing forms the siphons become elongated and reach up to the sand surface (Figure 11.1*e*). Other bivalves may bore through hard substances, using the anterior edges of the shell valves aided by chemical secretions. *Teredo* the ship-worm bores through wood, and the chalk-borer *Pholas* is credited with having excavated the English Channel. Some bivalves even swim: for example the scallop *Pecten*, which has eyes all round the mantle edge, can take off and swim by clapping its shell valves together, expelling a jet of water.

11.3 How do bivalves feed?

Ciliary feeding by means of the ctenidia is characteristic, but probably the earliest bivalves fed like the protobranchs, deposit feeders collecting food particles with the labial palps (e.g. *Nucula* (Figure 11.2*a*). A few protobranchs such as *Solemya* rely for food on chemosynthetic bacteria in the gills, with reduction or loss of the labial palps and part of the gut. The vast majority of bivalves, however, are 'lamellibranch': the name denotes not a division of the Class but a condition of gill enlargement and elaboration, achieved many times separately within a number of families. The free edge of the lamella of each ctenidium is turned up and attached to the mantle or to the foot so that it becomes a four-fold compacted structure (Figure 11.2*b,c*). The gill surfaces become pleated, making grooves bearing two main ciliary tracts at right angles: the lateral cilia produce a water current and the frontal cilia trap food particles (aided by the long laterofrontal cilia) and beat them (in mucus secreted by the gill) to a ciliated anteroposterior groove, either at the top or at the free edge (Figure 11.2c,d). The food is then conveyed forward to the labial palps, where large particles drop down grooves and are rejected while small particles pass on to the mouth.

11.3.1 Digestion

Digestion is largely intracellular. The stomach is modified for the continuous arrival of small food particles bound together in mucous strings. A 'style sac' contains the 'crystalline style', a rod made of compacted mucus containing amylase and cellulase enzymes. As the style rotates by ciliary action against a chitinous 'gastric shield', the tip is continually worn away, with release of a small quantity of enzymes. The rotation also mixes the stomach contents and pulls in more mucous strings of food.

11.3.2 Septibranchs

These are a small group of carnivorous bivalves that have abandoned ciliary feeding. The ctenidium has been lost and replaced by a muscular pumping diaphragm that draws in water, carrying prey (see Figure 11.2b). This is a bizarre reversal of the main direction of bivalve evolution, especially as it is combined with the crystalline style method of digestion, where there can be no extracellular protease. There are two deep sea genera, *Cuspidaria* and *Poromya*.

11.4 What kinds of muscle are there in bivalves?

Bivalves need to maintain muscle tension at low metabolic cost, because the two halves of the shell need actively to be held together: when the adductor muscles relax, the elastic recoil of the hinge ligament opens the shell. Where there is a byssus, its retractors must similarly maintain tension (see Figure 11.1). This need is met by a special form of muscle made of large 'paramyosin' filaments. It is slow to contract, inelastic and very isotonic, i.e. capable of maintaining tension at very different lengths (see Box 5.1). While it is very economical, the description of it as 'catch muscle' (implying that it can lock solid)

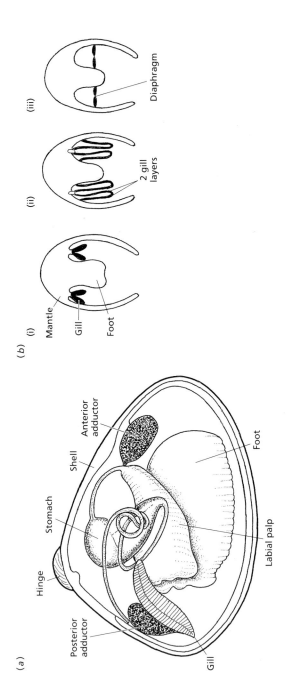

(a)

Hinge

Posterior
adductor

Gill

Stomach

Shell

Anterior
adductor

Foot

Labial palp

(b)

(i)

Mantle

Gill

Foot

(ii)

2 gill
layers

(iii)

Diaphragm

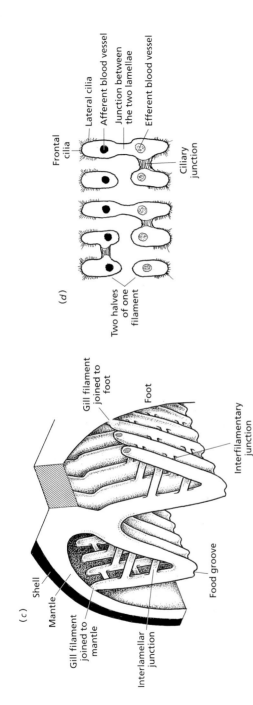

Figure 11.2. The functions of gills in bivalves: (a) *Nucula*, a protobranch, showing the large feeding palps and the small respiratory gills; (b) diagrams of bivalves in vertical section (V.S.), showing different degrees of gill elaboration, (i) primitive condition, (ii) lamellibranch, (iii) Septibranch; (c) V.S. lamellibranch gill, a stereogram showing the double layers; (d) transverse section of double-layered gill (less compacted than the one shown in (c).

(c)

Shell

Mantle

Gill filament joined to mantle

Interlamellar junction

Gill filament joined to foot

Foot

Interfilamentary junction

Food groove

(d)

Frontal cilia

Lateral cilia

Afferent blood vessel

Junction between the two lamellae

Efferent blood vessel

Ciliary junction

Two halves of one filament

is misleading because contraction does need to be maintained by occasional nerve impulses.

Bivalves also contain the normal invertebrate helical smooth muscle and occasionally fast-contracting striated muscle. The adductor muscle of the scallop *Pecten*, for example, can be seen to contain different regions: the more translucent is 75% striated muscle, used for quick twitches when swimming, and the more opaque region with long filaments is 56% paramyosin, used for holding the shell valves together.

SCAPHOPODA

Scaphopoda comprise a small (400 species) and very uniform group of marine molluscs encased in a one-piece tubular shell (Figure 11.3*a*). One shallow water genus, *Dentalium*, was much used for early embryological work. Scaphopods probably share ancestry with bivalves, but are sufficiently unlike other molluscs to be placed in a separate Class. The mantle, together with the mantle cavity, surrounds the whole animal and secretes the tubular shell, leaving openings at both ends. Ctenidia have been lost; water flows in and out of the upper (posterior) end and respiratory exchange occurs in the mantle.The elongated foot emerges from the larger (anterior) aperture and is used for burrowing. The proboscis-like head bears a radula and clusters of contractile tentacles that collect small particles of food (Figure 11.3*b*).

CEPHALOPODA

These very remarkable molluscs have become jet-propelled predators in the sea, with the evolution of highly developed eyes, brains and behaviour. They are the only invertebrates to have filled the same ecological niche as fish, and they continue to fill it in competition with fish: squids too may be very numerous and form fast-swimming shoals. Cephalopods also include the largest invertebrates: the body of a giant squid may be 10 metres long.

11.5 How is the molluscan body plan modified in cephalopods?

Whether the shell is retained, as in *Nautilus*, or internal and reduced or lost, as in coleoids, the animals are expanded along the dorsoventral axis and at the

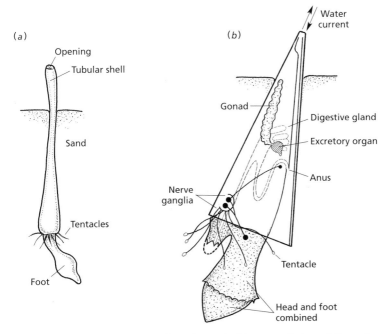

Figure 11.3. The structure of scaphopods: (*a*) external view of *Dentalium*; (*b*) diagrammatic longitudinal section of *Dentalium*.

same time the ventral side becomes anterior (Figure 11.4*a*). The mantle cavity is ventral and opens anteriorly (behind the head). Part of this opening is filled by the funnel, which represents part of the foot, through which water is expelled in jet propulsion (see below). The ctenidia lack cilia, being ventilated by the stream of water drawn through the mantle cavity by contractions of the mantle wall. There is a closed blood circulation in place of an open haemocoel. The foot is represented by the funnel and by the muscular tentacles (arms) round the mouth. Prey is seized by the arms, masticated by beak-like jaws and transported further by the radula. Eyes are large and prominent.

11.6 What cephalopods are known?

11.6.1 Ammonites

These well-known and commonly found fossils are among a variety of extinct groups with external shells (Figure 11.4*b*). They became extinct at the end of the Cretaceous, as did the dinosaurs.

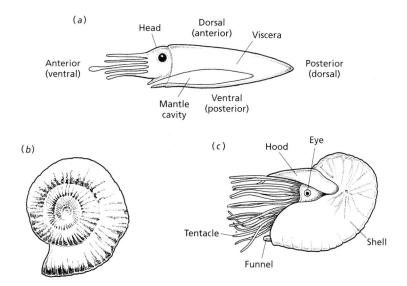

Figure 11.4. External views of cephalopods: (*a*) orientation of the cephalopod body (the equivalents in other molluscs are given in parentheses); (*b*) a fossil ammonite; (*c*) *Nautilus*;

11.6.2 Nautiloids

The nautiloid group also have external shells but are not closely related to the ammonites. *Nautilus* is a present-day genus (with five species) that retains the external coiled shell, divided by septa into compartments. The animal lives in the open final compartment, with a thread of tissue the 'siphuncle' extending back to the apex of the spiral. It is central in position, unlike the marginal siphuncle of ammonites.

11.6.3 Coleoids

These highly modified cephalopods (about 700 species) are without external shells (Figs. 11.4*d,e,f*). They include:

– **Decapods** with ten arms, such as the cuttlefish *Sepia* with an internal shell and squids such as *Loligo* with an internal horny rod in place of a shell.
– **Octopods**, with eight arms, have no shell at all.

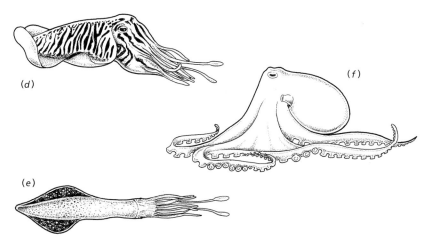

Figure 11.4 (cont.). (d) *Sepia*, a cuttlefish; (e) *Loligo*, a squid; (f) *Octopus*.

11.7 How is *Nautilus* able to survive?

Since, like many extinct cephalopods, it retains an external shell, *Nautilus* has been regarded as a deep water relict, surviving by chance. Recent study reveals it to be a specialised modern animal with unusual adaptations (and that it is probably more closely related to the coleoids than to the ammonites).

11.7.1 Buoyancy control

Control of buoyancy is an important key to its success, as *Nautilus* can readily maintain or alter its position in the sea (see Box 11.1).

11.7.2 Metabolic economy

A low metabolic rate is a striking attribute of *Nautilus*. It is an opportunist feeder, seizing crustaceans or small fish, and burns up its food unusually slowly. It can reduce this metabolic rate even further when starved and can continue to function at low oxygen tensions. *Nautilus* needs to live at low temperatures, below 20 °C. When very cold, the animal withdraws into the shell and remains motionless, building up an oxygen debt. A slow pace of life is, in general, characteristic of *Nautilus*. Reproduction is by slow direct development in particularly large eggs. Further development is also slow, and *Nautilus* appears to live for longer than other cephalopods.

Box 11.1
Buoyancy

A small animal will float at the surface of water; it is said to be neutrally buoyant. Most animals, however, are heavier than water ('negatively buoyant') and tend to sink, especially in fresh water, which contains very little salt. An animal that must swim continuously in search of food or in order to ventilate its gills can correct the tendency to sink without extra energy cost, but this is rare. A large cuttlefish, for example, would need 10% more energy if it had no float. Most marine animals have adaptations enabling neutral buoyancy: jellyfish contain fewer heavy sulphate ions and more of the lighter chloride ions than the surrounding sea; siphonophores and various other animals have gas-filled floats; some squids have sacs storing ammonia; pelagic molluscs lose or reduce their shells; arthropods lighten their skeletons; and there are very many other examples.

Two groups of animals go further, having adjustable mechanisms giving them neutral buoyancy at different depths in the sea. These are those teleost fish that have gas-filled swimbladders in which the pressure can be adjusted to match the hydrostatic pressure at different depths, and the cephalopod molluscs. The compartments of the shell, external in *Nautilus* and internal as the cuttlebone in cuttlefish such as *Sepia*, contain an adjustable mixture of gas and sea water. These cephalopods can maintain or alter their position in the sea with very little expenditure of energy; indeed, the ability to secrete gas into the shell may have been a crucial first step in the evolution of cephalopods. The shell is much thicker than in the ammonites, enabling *Nautilus* to live at depths down to 600 metres in tropical seas. Off the coral reefs of more temperate seas in the South Pacific, it is found at about 350 metres by day, where it is safe from turtles and other predators. At night it ascends to 150 metres, a level where there is more food. Buoyancy control is achieved by altering the proportions of gas and water in the shell compartments (see diagram (a)). When first formed, the chambers are full of sea water. The siphuncle takes up ions from the sea water by active transport (see Box 7.1), water follows by osmosis and enters the blood space within the siphuncular tube (see diagram (b)) to be swept down into the animal. In the chambers, gas bubbles out of solution, replacing the water removed. The animal will then be more buoyant and float higher in the sea. When the siphuncle pumps ions in the opposite direction, from the blood to the chamber, water will follow and there will be less gas in the chambers; the animal will therefore be less buoyant and sink. In cuttlefish the chambers of the internal shell are lined by a membrane functioning very much as does the siphuncle in *Nautilus* and buoyancy is regulated in much the same way.

Many fish also undergo daily vertical migration. They alter the pressure of gas in the swimbladder, rather than the proportions of gas and water; this process requires more energy and cannot give neutral buoyancy as quickly or as completely as is possible in *Nautilus*.

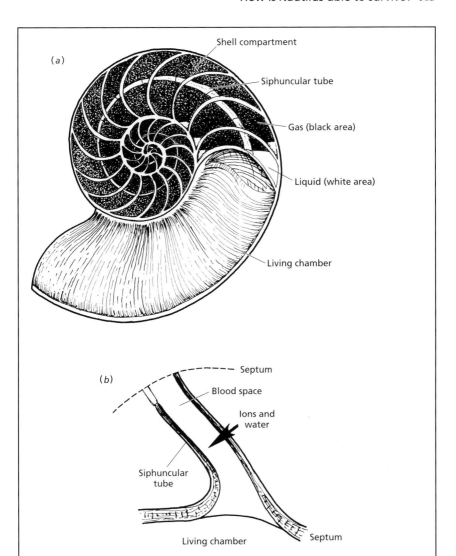

(a)

Shell compartment

Siphuncular tube

Gas (black area)

Liquid (white area)

Living chamber

(b)

Septum

Blood space

Ions and water

Siphuncular tube

Living chamber

Septum

The control of buoyancy in *Nautilus*: (a) gas and water in the shell compartments; (b) the siphuncular tube enlarged, with an arrow indicating the direction of movements of ions and hence water, causing an increase in the proportion of gas.

11.8 How have some cephalopods become so active?

11.8.1 Loss of the external shell

This allows great mobility: armour has been sacrificed to speed. Jet propulsion is achieved by strongly developed circular muscles round the mantle. Powerful contraction expels a fast jet of water, directed by the funnel: contraction of vertical muscle within the mantle wall allows water to enter and the mantle cavity is re-extended. In addition to the funnel jet there may be fins, which not only steady the cuttlefish and squids but by undulating waves provide the main method of slow movement.

11.8.2 Buoyancy

Cuttlefish are buoyant, using a mixture of gas and air in the compartments of the internal shell (see Box 11.1). Squids have no shell (merely a stiffening rod along the back) but may increase their buoyancy by retaining ammonia in special reservoirs (ammonium chloride is lighter than the sodium ions in sea water). Octopuses are not buoyant, they are bottom-living animals.

11.8.3 Defence

It is important to have some form of defence when there is no external shell. The sudden production of an ink screen can hide the animal. Coleoids also can become camouflaged by changing colour: they have 'chromatophores', which are elastic sacs containing pigments, usually blue, orange and yellow. These sacs can be flattened by contraction of a ring of radially arranged muscles so expanding a disc of colour; when the muscles relax the sacs become smaller and spherical and the colour spots contract. Colour change is under nervous control and is very rapid. This is in contrast to most invertebrate chromatophores where the pigments are dispersed through branching processes or concentrated in the central body of the cells; these pigment movements are typically under hormonal control and occur relatively slowly.

11.8.4 Feeding

Feeding is active and rapid. The arms are covered in strong suckers: decapods extend two long arms to seize their prey; octopuses jump on to a crab (or whatever the prey may be), entangling it in the interbrachial web (Figure 11.4*f*). Toxins are secreted as the jaws bite.

11.8.5 Circulation of the blood

Closed circulation can be much more efficient with capillary vessels between the arteries and veins. The blood pressure is relatively high (for a mollusc) and may be boosted by 'branchial hearts' at the bases of the gills.

The most remarkable evolutionary advance in cephalopods is the development of the sense organs, nervous system and brain.

11.8.6 Sense organs

The suckers on the arms serve also as sensory receptors, the statocysts beside the brain detect gravity and changes in angular acceleration and changes in water displacement are detected by a lateral line system, a remarkable parallel with fish. Rows of hairs parallel to the long axis of the visceral hump are continued over the head and along the arms: these receptors connect to sensory nerves, where the frequency of nerve impulses depends on the frequency and intensity of water movements. The most striking resemblance to vertebrates is shown by the eyes (see Figure 11.5*a*). Apart from the external origins of the optic nerve, this could be the eye of a fish.

11.8.7 The brain

The cephalopod brain is the largest and most complex among invertebrates. Correlated with the sophisticated sensory equipment and complicated behaviour, the basic molluscan brain (a ring of nerve around the oesophagus) has become elaborated to form an organ that can be compared with that of vertebrates: indeed the brain to body weight ratio in a coleoid is greater than that of many fish and reptiles. The sea-slug *Aplysia* has some 20 000 nerve cells divided into 10 groups: the octopus brain contains an estimated 520 million, arranged in about a dozen separate lobes (see Figure 11.5*b,c*), each with particular functions. Lobes concerned with motor output are ventral to the oesophagus; dorsal to it are lobes processing the various kinds of sensory input (touch, sight, etc.). The highest centres (absent in *Nautilus*) concern learning and long-term memory.

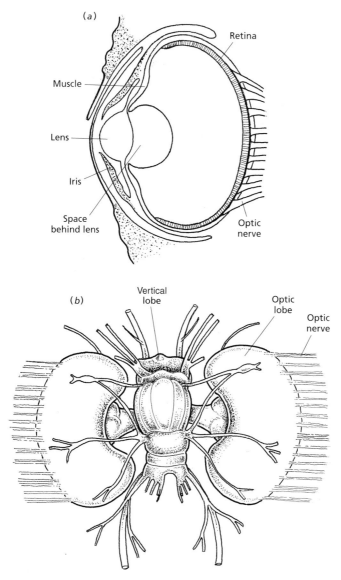

Figure 11.5. (a) The eye of a cephalopod; (b) the brain of an octopus in dorsal view;

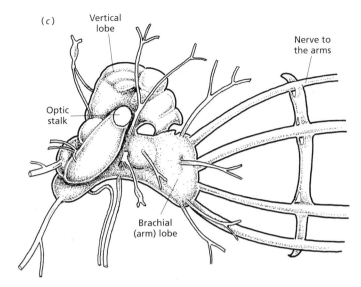

(c)

Vertical lobe

Nerve to the arms

Optic stalk

Brachial (arm) lobe

Figure 11.5 (*cont.*). (c) the brain of an octopus in lateral view.

11.8.8 The nerves

The nervous system is correspondingly well-developed. In *Loligo* the stellate ganglion (about 120 000 neurons) gives off branches all over the mantle, producing simultaneous muscle contraction. The giant nerve fibres, as much as a millimetre in diameter, conduct extremely quickly. Experiments on these nerves were the source of much of our knowledge about nerve action in general (see Box 11.2). With this equipment, the great learning power of coleoids is no surprise.

11.8.9 Reproduction

Reproduction has unexpected features for animals so advanced. The eggs may be brooded, as in the octopus, but the young never meet their parents. Development is direct in a large yolky egg (molluscs so mobile do not need larvae for dispersal). The behaviour of a newly hatched cephalopod is very stereotyped, and only later shows the flexibility that the brain allows. They live for one or two years, have a reproductive burst and then die.

Box 11.2
Nerves and brains

Neurons (or neurones)

Most animals are coordinated by nervous systems. These consist of separate neurons that transmit information by conducting nerve impulses. Neurons can conduct impulses over long distances owing to their characteristic thread-like shape: typically there is a long fibre, the axon, leading from the nucleated cell body, which also bears shorter processes, the dendrites (see p. 170). Ganglia are groups of neurone cell bodies; they include interneurons, which may or may not have axons but make connections between other neurons.

Synapses

The gaps at the ends of axons and dendrites are called synapses. They conduct in one direction only, providing information to other nerve cell bodies or to other organs such as muscles. Transmission across neuromuscular junctions is usually chemical, often by means of acetylcholine, and after transmission the chemical is rapidly destroyed. Synapses within the nervous system may use chemical or electrical transmission; they may receive input from more than one neuron and may be either excitatory or inhibitory in effect.

Brains

Most animals have a coordinating nerve centre or brain, usually near the front end, to which 'afferent' nerves carry impulses from the sense organs and from which 'efferent' nerves carry impulses to muscles, glands or other 'effector' organs. Different animals have elaborated their brains to different extents according to their requirements.

Neurosecretion

Many neurons are specialised for chemical secretion. For example, while vertebrate hormones are usually secreted into the blood by ductless (endocrine) glands, often under nerve control, invertebrates usually secrete hormones directly from nerve cells into the blood or tissue fluid. Neurosecretory cells have large nuclei and axons with terminal swellings closely applied to fluid spaces. Moulting in arthropods (see Chapter 12) provides examples.

Invertebrate nervous systems

Sponges have no nervous systems. Cnidaria and echinoderms have simple nervous systems without brains. Most unsegmented worms have bilateral symmetry with some anterior sense organs and a brain that is little more than a sensory motor link. Cell bodies and axons usually form about two main lateral or ventral nerve cords.

Annelids

Annelids have anterior ganglia above and below the oesophagus, linked by nerve loops called commissures. The supra-oesophageal ganglion may be called the brain; it provides simple coordination between sensory input and motor output, and in some annelids (including earthworms) is not even needed to initiate movement. The suboesophageal ganglion is the first of a chain of segmental ganglia along the double ventral nerve cord (see page 170). Behaviour is largely based on segmental reflexes. Giant fibres with rapid conduction are important in burrowing worms and in tube-living polychaetes. Leeches have more specialised nervous systems, with the segmental ganglia condensed together and large ganglia under the suckers.

Molluscs

No generalisation can be made. Gastropods typically have an anterior 'cerebral' ganglion with very few controlling functions and about three other main ganglia, serving the viscera, the mantle and the foot. As well as nerve fibres connecting these ganglia there are longitudinal nerve cords, often twisted into a figure of eight by torsion. Sessile bivalve molluscs and the very active cephalopods represent extremes of brain development within one phylum (see Chapter 11).

Arthropods

Arthropod nervous systems are built on the same plan as those of annelids (see page 170) but arthropod brains are usually highly developed and form the basis for most elaborate behaviour (see Chapters 12–15).

Chordates

Unlike other phyla, chordates have hollow dorsal nerve cords, though in invertebrate chordates these may only occur in the larval stages (see Chapter 18).

Box 11.2 (*continued*)

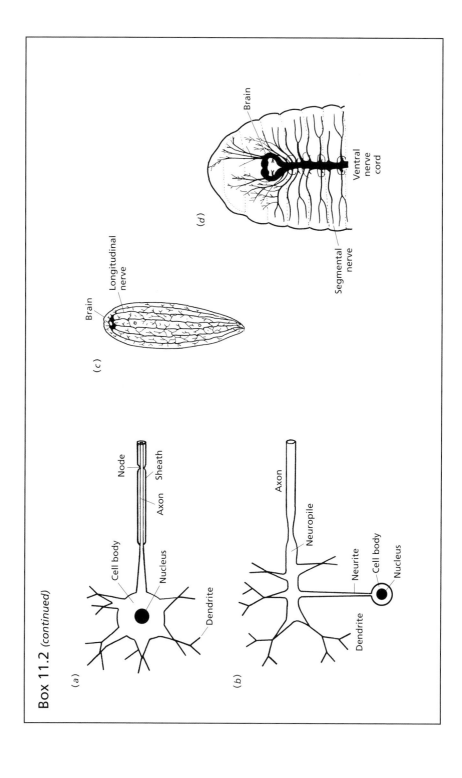

(a) Cell body · Node · Sheath · Axon · Nucleus · Dendrite

(b) Axon · Neuropile · Dendrite · Neurite · Cell body · Nucleus

(c) Brain · Longitudinal nerve

(d) Brain · Segmental nerve · Ventral nerve cord

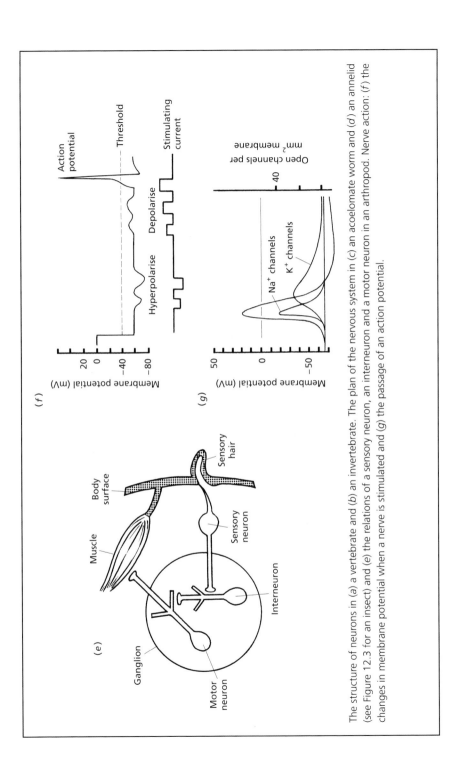

The structure of neurons in (a) a vertebrate and (b) an invertebrate. The plan of the nervous system in (c) an acoelomate worm and (d) an annelid (see Figure 12.3 for an insect) and (e) the relations of a sensory neuron, an interneuron and a motor neuron in an arthropod. Nerve action: (f) the changes in membrane potential when a nerve is stimulated and (g) the passage of an action potential.

Box 11.2 *(continued)*

How nerves work

The basic facts are as follows:

1. Electric signalling along nerves is achieved by charged ions moving in and out of the axon (contrast the flow of electrons along a telegraph wire)
2. Passive ion movement is determined by chemical and electrical gradients
3. Axon membranes are selectively permeable to sodium and potassium ions. There are separate sodium and potassium channels, each regulating the passive flow of the relevant ion.
4. Axon membranes, like the membranes of nearly all cells, have in addition a 'sodium pump': an active (energy-requiring) process that transports sodium ions outwards, resulting in more sodium outside cells and more potassium inside them.

Resting nerve

The cell membrane is impermeable to the large internal protein molecules, which are predominantly negatively charged. Ions move freely in and out until equilibrium is reached. Due primarily to positively charged potassium ions moving out, the result is a potential difference of about 75 millivolts (mV) across the membrane, with the outside positively and the inside negatively charged. Sodium ions stay outside owing to the outward sodium pump. The membrane is said to be **polarised**. Decrease in the resting potential is called **depolarisation** and increase is called **hyperpolarisation**.

Nerve impulses

Impulses arise and are propagated when the resting nerve is stimulated. A patch becomes permeable to sodium ions which rush in, not only depolarising the membrane but reversing the polarisation; the inside becomes positively charged (with a potential difference of about 55 mV). This is quickly reversed again by increased outflow of potassium, repolarising the active patch while the impulse is propagated along the nerve (see diagram g).

This process is very economical, unlike muscle contraction. It requires only a very small amount of energy to restore the ion pools.

11.9 What has limited the evolution of cephalopods?

How is it that animals so highly developed as active predators have not become even more successful? Part of the answer must be that they are limited by their molluscan constitution. For example, digestion is slow in the cuttlefish and the octopus, where the digestive gland is used both for enzyme secretion and for food absorption. Digestion of a meal takes 24 hours. In squids, food is absorbed in the caecum also and the time is cut to 4 hours or less. The circulatory system even with its modifications may not supply oxygen to muscles very fast. Basic molluscan structure may be limiting in that flexible arms, with no joints as fixed reference points, cannot discriminate between shapes, however elaborate the brain: an octopus can have only limited powers of manipulation. Perhaps most importantly, the kidney remains that of a primitive mollusc, the blood pressure is not high enough for rapid filtration and the skin is too specialised to assist active ion transport or to keep water out. Cephalopods accordingly have negligible powers of osmo-regulation and at the same time are too complex to tolerate changes in internal concentration. They are confined to the sea.

Perhaps we, descendants of the early vertebrates, should be relieved that there were limits to the performance of these extraordinary molluscs and that their degree of specialisation has prevented them from evolving further.

12

Arthropoda: general

Arthropods are segmented animals with an outer cuticle, the exoskeleton, which is typically hardened and inflexible over much of the body but remains flexible at joints; the muscles are attached inside this cuticle. Like a molluscan shell it protects the soft parts, but, unlike a shell, the arthropod cuticle is built into the animal. With jointed limbs, fast-contracting striated muscle and a well-developed nervous system, rapid locomotion can be achieved.

The arthropod body plan has proved outstandingly successful. Arthropods constitute nearly nine-tenths of all known animal species, occurring in the sea, in fresh water, on land and in the air, in every conceivable ecological niche. Not only are they many and diverse as species (at least a million have been described) but also they are numerous as individuals, for example there are about 10^{21} copepod crustaceans in the oceans and some 200 million insects are said to exist for every human being on earth.

The outstanding success of arthropods is largely attributable to the nature of the cuticle. This is the primary theme of this introduction to the arthropods. The diversity of arthropods is described in the three following chapters, with emphasis on evolution from the sea to fresh water and land. Chapter 13 introduces the Crustacea, which are mainly aquatic. Chapter 14 includes Chelicerata, the marine horseshoe crabs and terrestrial arachnids, and also Myriapoda, all terrestrial. Chapter 15 discusses Insecta, which are nearly all terrestrial as adults, and often aerial.

The evolutionary relationships of arthropods constitute a further theme: originally they were assumed to belong to a single phylum, derived from segmented annelid-like coelomate worms. However, both the origins of arthropods and the relationships between them have been very controversial. Are they a single phylum or has this type of exoskeleton evolved several times separately? Most of their defining characteristics (such as jointed limbs, the need to moult, etc.) are necessary consequences of having a hard exoskeleton: for example, should crustaceans, insects and chelicerates be placed in different phyla? Modern evidence increasingly confirms the original assumption of monophyly. Chapter 20 combines fossil, molecular and morphological evidence to discuss arthropod relationships.

12.1 What defines an arthropod?

The exoskeleton covers the whole body. The body segments are often grouped into regions, for example head, thorax and abdomen. The most anterior segment never bears appendages, i.e. the anterior region has at least one more segment than the number of paired appendages. Primitively, probably each segment behind the most anterior bore paired appendages but in most groups many have been lost, as in many parasites and in the posterior regions of insects.

The cuticle is moulted at intervals when the young animal increases in size; the process is controlled by hormones. The brain, sense organs and nervous system are extremely well developed: rapid and precisely controlled movement is characteristic of arthropods and complex behaviour has evolved. The coelom is much reduced (it may be represented only in the gonads) and the main body cavity is a haemocoel containing blood, which enters the dorsally placed heart through holes (ostia). There are no locomotor cilia, but some sense organs are constructed from ciliary structures. As will be explained, respiratory and excretory organs differ greatly between aquatic and land forms. Typically the sexes are separate. The need for internal fertilisation is often associated with complex behaviour. Development is basically protostome (see Box 5.3) but much modified in most arthropods.

12.2 What are the key features of arthropod cuticle?

The frequent statement that 'to study cuticle is to study 90% of the properties of an insect' gives an emphasis that can be extended to all arthropods.

12.2.1 The cuticle

The cuticle or exoskeleton is made of chitin and protein. Chitin is a nitrogenous polysaccharide, related to mucus from which perhaps it was derived. It

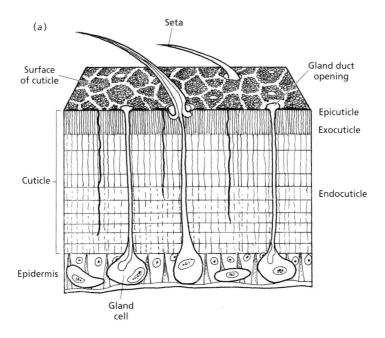

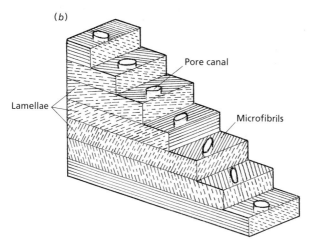

Figure 12.1. Arthropod cuticle: (a) diagrammatic section through arthropod cuticle; (b) diagrammatic section through a section of cuticle showing the progressive change in orientation of the chitin–protein microfibrils in successive lamellae of the exocuticle, forming a helicoid structure.

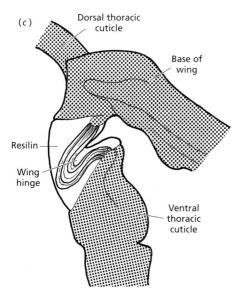

(c)

Dorsal thoracic cuticle

Base of wing

Resilin

Wing hinge

Ventral thoracic cuticle

Figure 12.1 (cont.). (c) Transverse section through the thoracic wall of *Schistocerca*, a locust, showing the pad of resilin at the wing hinge.

is not in itself hard: microfibrils of chitin are embedded in protein and the cuticle acquires strength and rigidity when the chains of protein molecules become cross-linked ('tanned') by phenols (quinones).

12.2.2 Cuticle is laid down by epidermal cells

The epidermal cells that lay down the cuticle are spread over the animal in a single-layered sheet. Each cell secretes its own overlying cuticle, giving precise spatial control.

12.2.3 Cuticle is laid down in layers

Low magnification reveals:

- **Epicuticle:** protein and wax; a protective outer surface (shed at moulting).
- **Exocuticle:** chitin and tanned protein, hard and often dark (shed at moulting).
- **Endocuticle:** chitin and soft protein; (reabsorbed at moulting).

The epidermal cells are underneath (Figure 12.1*a*). The layers are crossed by ducts of dermal glands and by pore canals, probably conducting wax. Electron microscopy reveals regular variations in the microstructure of the many layers constituting the exo- and endocuticles. Microfibrils of chitin and protein are orientated according to the mechanical demands on particular regions: most typically they change direction systematically through the thickness of the cuticle to form a helicoid (Figure 12.1*b*), an arrangement conferring strength and resistance to cracks.

12.2.4 Cuticle is extremely versatile

As well as supporting the animal and attaching muscles, parts of the cuticle need to be modified for special functions; for example, it is very hard and rigid over claws and biting mouthparts but thin and flexible at joints. It is capable of deformation where required, with elastic properties that enable recoil and storage of energy. It needs to protect the animal while at the same time admitting information from the environment. To achieve all this requires a formidably complex composite material.

12.2.5 Lipid

Lipid (wax) in the epicuticle forms the main barrier to water loss in terrestrial arthropods. Waterproofing with wax avoids the use of heavy materials. The wax layer is delicate: tanned proteins in the cuticle below prevent deformation, but an insect cannot for instance go in and out of soil without abrading the epicuticle unless the wax has become protected by cement. The surface lipids serve also to deter predators and parasites, control temperature to some extent and deflect unwanted water. They may convey chemical signals to mates: the great variety of lipids in insect epicuticles suggests their use as pheromones (hormones acting outside the body). Epicuticular lipids are probably crucial to the survival of most arthropods.

12.2.6 Resilin

A uniquely elastic protein, resilin is capable of complete recovery of its shape after deformation. It consists of polypeptide coils linked at their sides by amino acids. Although the elastic properties of resilin (Figure 12.1*c*) are especially vital to flying insects, resilin is also used by other arthropods, for example in the claws of scorpions, where there is no extensor muscle.

12.2.7 Modifications between groups

Different arthropod groups show characteristic modifications. Crustacea, mostly living in water where weight is not limiting, commonly have cuticles further hardened by calcium salts. In insects and arachnids where water conservation is particularly important, the wax content of the epicuticle varies according to habitat, and can be altered by the animal in different conditions.

12.2.8 Moulting

Understanding of cuticle demands study of its formation in time as well as in space; study of moulting is revealing. To some extent arthropods can change in size and shape by making their exoskeleton more flexible, but once it is tanned, cuticle cannot be expanded. Moulting is dangerous, as the animal is briefly incapacitated and vulnerable, and wasteful of material. But larvae and their component structures such as mouthparts must increase in size, and this requires the shedding and replacement of the cuticle. The sequence of changes at moulting is shown in Figure 12.2.

These changes culminate in ecdysis, the shedding of the old cuticle. The arthropod swallows some of the medium (water or air), pressure from inside splits the old exocuticle and the animal wriggles out. It quickly expands and tans the proteins in the outer layer of the new cuticle. The living tissues of the animal then grow and the cuticle is fully reconstructed.

The need to moult during growth provides an opportunity for 'metamorphosis', an abrupt change in form during the life cycle, important in many arthropods. Moulting and metamorphosis are under hormonal control. Figure 12.3 indicates the series of steps in insects and in crustaceans such as crabs, crayfish and prawns ('decapods', see Chapter 13). The brain initiates neurosecretion of a hormone that is released into the blood from a structure near the brain (the corpus cardiacum in insects, the sinus gland within the eye stalk of crustaceans). In insects the brain hormone activates production of a moulting hormone precursor from the prothoracic gland, in decapod crustaceans the brain hormone inhibits hormone production by the Y-organ. When the ecdysone precursor reaches the epidermis, ecdysone is formed and stimulates the cells to divide. Superimposed on these similarities are important

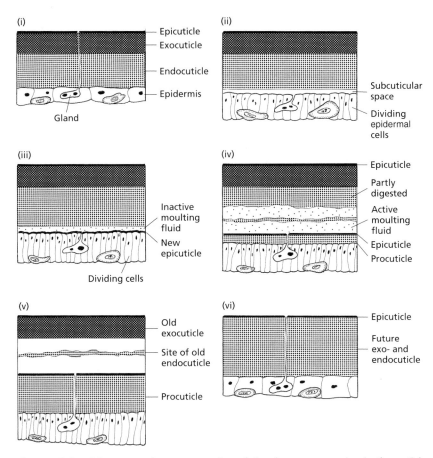

Figure 12.2. Diagrammatic representation of the changes occurring in the cuticle during the moulting cycle: (i) mature cuticle; (ii) epidermis divides; (iii) new epicuticle layer produced; (iv) endocuticle digested; (v) moulting fluid resorbed; (vi) remains of old cuticle has been cast off.

differences: unlike many crustaceans, insects (except mayflies) never moult as adults but have a 'juvenile hormone' produced in a separate group of cells (the 'corpus allatum'). It diffuses to the epidermis and, at every moult except the last, promotes the appearance of larval form.

12.3 How are arthropod internal cavities organised?

The organisation of the soft parts of an arthropod is dictated largely by the cuticle: for example, the absence of cilia is not surprising, since they could not work on the outside and are not needed on the inside.

12.3.1 The haemocoel

With no requirement for a hydrostatic skeleton the coelom is greatly reduced. It is represented only in the cavity of the gonads and, in some aquatic arthropods, in that of the excretory organs. The main body cavity is the haemocoel, divided into blood-filled spaces that bathe all the tissues (Figure 12.4a). The blood is moved by muscular contraction. It enters the dorsally placed heart through the ostia and is pumped forward into vessels opening into the haemocoel; the heart beat is under nervous control in arthropods. This open system allows blood to be moved in bulk, which is important at moulting, providing turgor and maintaining the shape of the animal immediately after ecdysis. As well as dissolved food, hormones and other substances, the blood contains many cells, which may be stationary on surfaces or actively moving, such as the lurking phagocytes that constitute the immune system. In insects, where there is no separate digestive gland, other cells in the blood may store and process food.

12.3.2 The gut

The gut (see Figure 12.4a) runs from end to end of an arthropod, with infolding of cuticle both at the mouth and at the anus. The midgut may receive secretions from digestive glands, and may be lined with permeable cuticle through which food is absorbed into the haemocoel. Gut structure varies according to the great variety of food and feeding methods.

12.4 What makes possible the great activity of arthropods?

12.4.1 The segmental plan

In arthropods, the segmental plan combined with the hard exoskeleton allows rapid and precisely controlled movement. Unlike the muscular body wall of a worm, the muscles are discrete units controlled by segmental nerves. They move the segmental appendages. The most anterior appendages are usually sensory (e.g. antennae), the next pairs are usually concerned with feeding (e.g. hard biting mouthparts) and more posteriorly there may be many pairs of walking or swimming limbs.

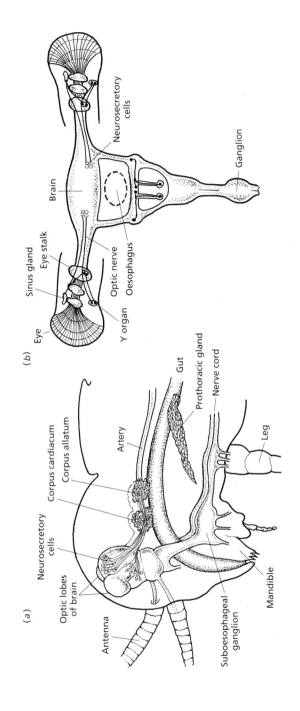

(a)

Neurosecretory cells

Corpus cardiacum

Corpus allatum

Optic lobes of brain

Antenna

Suboesophageal ganglion

Artery

Gut

Prothoracic gland

Nerve cord

Leg

Mandible

(b)

Eye

Sinus gland

Eye stalk

Y organ

Optic nerve

Oesophagus

Brain

Neurosecretory cells

Ganglion

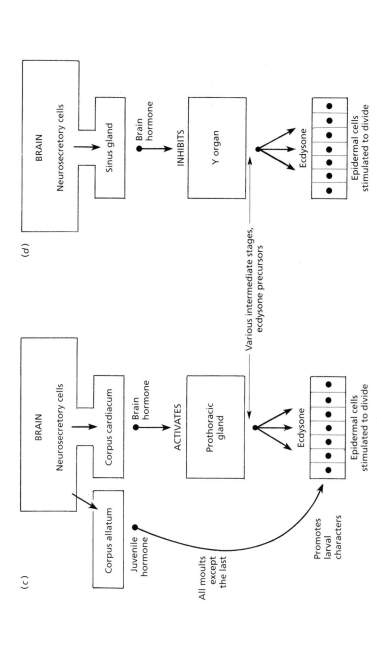

Figure 12.3. Hormonal control of moulting: (*a*) diagrammatic longitudinal section of the anterior end of an insect, to show the brain and other sources of hormones controlling moulting; (*b*) diagrammatic dorsal view of the eyestalks and the anterior part of the nervous system in a decapod crustacean; (*c*) insect and (*d*) decapod crustacean, comparison of control of moulting.

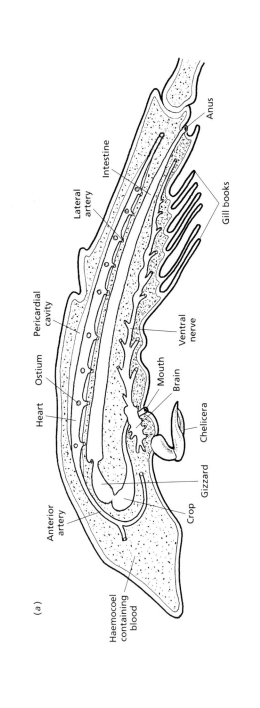

(a)

Haemocoel
containing
blood

Anterior
artery

Heart   Ostium

Pericardial
cavity

Lateral
artery

Intestine

Anus

Gill books

Ventral
nerve

Mouth

Brain

Chelicera

Gizzard

Crop

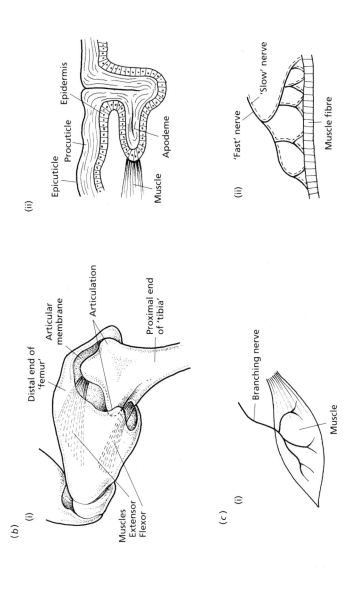

Figure 12.4. (a) A generalised chelicerate showing the open blood system; (b) arthropod (insect) leg joint, (i) showing muscle attachments and (ii) enlargement showing apodeme; (c) nerve control of muscle in arthropods, (i) nerve branching over the entire muscle and (ii) two different kinds of nerve with multiterminal twiglet endings over a single muscle fibre.

12.4.2 The brain

This organ is greatly elaborated from the simple sensory/motor exchange found in the ganglia of many invertebrates. Behind it there is a nerve ring round the gut and a double ventral nerve cord with a ganglion in each segment receiving sensory innervation and giving off motor nerves, much as in annelids. Head development in arthropods is advanced, and the brain is an organ initiating and controlling behaviour. It is dorsal to the anterior end of the gut and divided into three lobes: the most dorsal receives sensory axons from the eyes, the middle lobe receives axons from the antennae (it is absent in chelicerates, which have no antennae) and the ventral lobe contains motor neurons giving off motor nerves (see Figure 12.3*a*). Behind the brain there is a double ventral nerve with a pair of ganglia in each segment.

12.4.3 The neuromuscular system

All the muscle is striated. Like vertebrates, arthropods can move very fast with great efficiency, but they control their movements very differently. Arthropod muscle is unfamiliar to us in that it works inside a tube of exoskeleton, to which it is attached directly (often to infolded ridges of the cuticle called 'apodemes') or indirectly through tendons. Attached on either side of a hinge or a joint, muscles can build up tension and effect contraction with very little change in length, i.e. they are isometric (Figure 12.4*b* and Box 5.1).

Nerve control of muscle is far less centralised than in vertebrates: arthropod muscle response is finely adjusted at the nerve endings. A vertebrate limb muscle receives a large number of identical motor axons, each ending in a motor unit, and each muscle fibre is part of a motor unit. Contraction depends on the number of motor units stimulated and the frequency of stimulation, and any inhibition is also centralised. Arthropods have no motor units; instead a limb muscle receives few axons, which branch to more than one fibre and branch further to form 'twiglet' endings all over the muscle fibre (Figure 12.4*c*). The axons are not all alike: there are fast and slow fibres (the descriptions apply to the muscle response) and inhibitory ones. Vertebrate striated muscle contraction is 'all or none': if nerve stimulation reaches a threshold an action potential is generated and spreads all over the muscle, causing it to contract. In arthropods, muscle contraction is not all or none and muscle action potentials are rare: local 'junction potentials' occur under the twiglet nerve endings, causing a degree of contraction roughly proportional to the local electrical change. Muscle response is thus graded according to the local junction potential, which in turn depends upon which axon has been stimulated, and whether or not it is inhibited at the site.

12.4.4 Sense organs: sensilla

An animal encased in armour needs structures sensitive to environmental events. The exoskeleton is locally modified into 'sensilla', in the form of hairs, bristles or pits that can receive chemical or mechanical stimuli and set up electrical changes in associated sensory neurons. Receptors detecting mechanical change are found at the base of appendages and elsewhere (Figure 12.5a,b). Chemoreceptors resemble them but the sensory nerve extends to the tip of the hair: they are concentrated anteriorly, usually on the antennae, which may also contain sound receptors. In addition to these 'exteroceptors', there are at the joints 'proprioceptors' giving the animal precise information about the localisation and extent of its own movements. This allows a degree of control quite impossible for a worm, which cannot know exactly how its body bends.

12.4.5 The eyes

Compound eyes are characteristic of arthropods and occur in all groups. They are composed of a large number (as many as 4000 in some insects) of light-receiving units called ommatidia (see Figure 12.5c). Each consists of a cylinder of elongated 'retinula' cells, with a small external lens (Figure 12.5d). The refractive index in the cylinder is graded so that the lens focuses light to the central axis or 'rhabdome', which is the much folded inner surface of the retinula cells. Here is situated the visual pigment, which responds to light by depolarisation (see Box 11.2), which stimulates neurons to convey information to the brain. The ommatidia are separated by pigment, movable so that more or less light can be admitted.

A compound eye suggests a mosaic, with each ommatidium sensitive to a particular spot on the whole visual field. Most importantly, however, each ommatidium receives some of the light falling on its neighbours, so that the visual fields of neighbouring ommatidia overlap. Such a structure is extremely sensitive to movement, over a large field of vision, and this is the great advantage of a compound eye. It also provides magnification, and owing to the

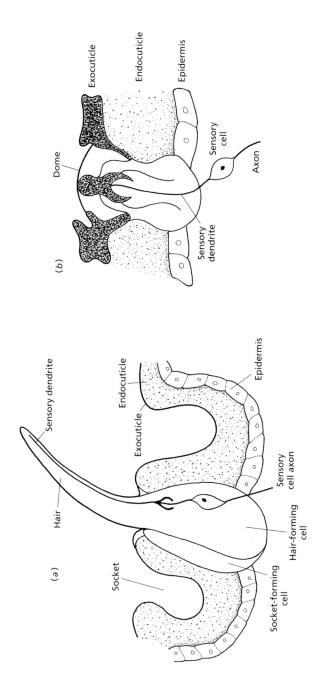

(a)

Sensory dendrite

Hair

Endocuticle

Exocuticle

Epidermis

Sensory
cell axon

Hair-forming
cell

Socket-forming
cell

Socket

(b)

Exocuticle

Endocuticle

Epidermis

Dome

Sensory
cell

Axon

Sensory
dendrite

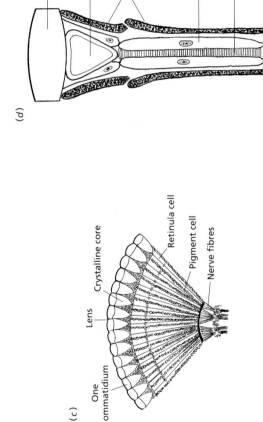

(d)

Lens

Crystalline core

Pigment cells

Retinula cell

Rhabdome

(c)

One ommatidium

Lens

Crystalline core

Retinula cell

Pigment cell

Nerve fibres

Figure 12.5. Arthropod sense organs: (a) insect mechanoreceptor with sensory hair; (b) Insect 'campaniform sensillum', a dome-topped structure detecting distortions in the nearby cuticle; (c) diagram of a compound eye; (d) diagrammatic longitudinal section of a single ommatidium.

short light path may readily detect short wavelengths, for example ultraviolet light. Compared to the vertebrate (or cephalopod) eye it, has however, poor resolution and image formation.

The above generalised discussion indicates that most of the characteristic features of arthropods relate to their exoskeleton, and begins to explain the foundation for their diversity, which will now be introduced.

12.5 What arthropod relations are known?

Fossils, in particular one large fossil group the Trilobites, are the outstanding omission from this account, but introduced in Chapter 20. **Crustacea, Chelicerata, Myriapoda** and **Insecta** are discussed in subsequent chapters; are there any other present-day arthropods? Two small groups have at times been included as 'arthropod-like', or 'protoarthropods': **Pentastoma**, now recognised as modified parasitic crustaceans (see Chapter 13) and **Pycnogonida**, now included as chelicerates (see Chapter 14).

Finally, to emphasise definition, there are two groups which are not arthropods but which in some ways resemble them:

1. **Tardigrada**. Despite some similarities such as the moulted cuticle, the tardigrades are not arthropods but constitute a separate phylum. These attractive little animals (Figure 12.6a) are the koalas of the invertebrate world. Some 600 species are known, only 0.1–1.2 mm long, living in water films on mosses or other plants in damp places or between particles in freshwater (less commonly marine) sediments. They have seven to eight segments, bearing short, lobe-like legs consisting mainly of claws, and a ventral nerve cord with segmental ganglia. Muscles radiate from the leg base and there is no circular muscle. These small animals have no blood vessels or segmental excretory organs. The body cavity is a pseudocoel and there are 'Malpighian tubules' regulating the ion content of the body fluid and opening into the gut. In dry conditions tardigrades survive by forming 'tuns', barrel-shaped resistant resting stages with greatly reduced metabolism; these may also be dispersal agents.

   Tardigrades probably evolved by miniaturisation, perhaps through early maturation of a small ancestral larva. What that ancestor might have been is uncertain, but they may be close to the lineage of arthropods, nematodes and some other animals with hard moulted cuticles.

2. **Onychophora**. Some 110 species of onychophorans are known, originally described as constituting a single genus, the 'velvet worm' *Peripatus* (body

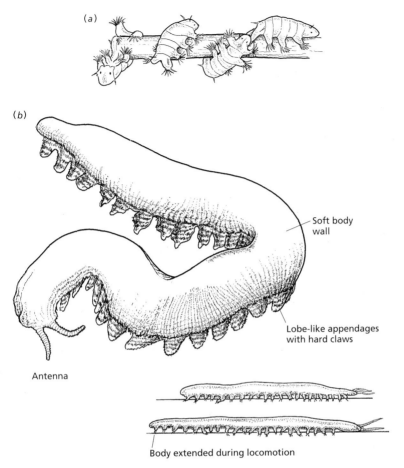

(a)

(b)

Soft body
wall

Lobe-like appendages
with hard claws

Antenna

Body extended during locomotion

Figure 12.6. Non-arthropods related to the phylum: (a) tardigrades; (b) *Peripatus*, an onychophoran.

length 14–150 mm). Living under damp logs, mostly in the countries of the southern hemisphere once united as Gondwanaland, they can vigorously squirt out mucus and look like slugs with legs: the soft body has no external signs of segmentation but is covered in soft (velvety) chitin, frequently moulted, with tubercles bearing iridescent scales. The body is, however, segmented (though without segmental nerve ganglia) with three pairs of head appendages – antennae, mandibles round the ventral mouth and oral papillae – followed by 13 to 43 pairs of walking legs, which are unjointed lobes terminating in claws (Figure 12.6*b*). This combination of characters led to *Peripatus* being hailed as a 'missing link' between annelids and arthropods (when arthropods were thought to have arisen from annelid-like ancestors). Closer examination shows that *Peripatus* is a specialised modern animal adapted to a particular way of life: the body wall is lined by muscles working against hydrostatic pressure in the body cavity, as in many worms, making it very compressible. The animal is able to squeeze under stones and logs and through openings as small as one-ninth of its resting diameter. Yet the jaws and claws of these predatory carnivores have chitin hardened by tanning of the associated protein, much as in arthropods. The body cavity is, as in arthropods, a haemocoel, the coelom being represented only in the cavities of the excretory and reproductive organs. Worm-like ciliated segmental tubules carry out excretion, yet respiration is achieved by arthropod-like tracheal tufts scattered all over the body. The spiracles cannot be closed and the cuticle is very permeable for a terrestrial animal, leading to great loss of water. Water absorption by the lobes between the legs is probably very important in enabling these animals to live on land, and they are confined to moist habitats. Reproduction is another problem. Remarkably, many species are viviparous, with developing eggs not only retained but supplied with food in the mother's body.

The recent discovery of fossil aquatic 'lobopods' suggests the ancestry of this ancient and isolated phylum, separate from arthropods but fairly close to their origin. Arthropod relationships are considered in Chapter 20 but first the main groups of present-day arthropods are discussed.

13

Crustacea

Crustacea include crabs, lobsters, crayfish, woodlice, barnacles and many kinds of shrimp. They are primitively and predominantly marine: of some 38 000 species known, 84% live in the sea, 13% live in fresh water and only 3% are terrestrial. They are very successful arthropods, being extremely numerous as individuals and also diverse: in length alone, crustacea range from less than a millimetre to 4 metres (in some spider crabs).

13.1 What is distinctive about crustaceans?

How can they be recognised among other animals with the general arthropod characters? The main diagnostic character is that crustaceans have two pairs of antennae; if you find an obvious arthropod, count its antennae. The first head segment has no appendages, the next two both bear antennae.

Typically but not always, the limbs are biramous (two-branched) and usually the head appendages include two pairs of maxillae.

Typically the life cycle includes a 'nauplius' stage, either as the earliest free – swimming larva with three pairs of appendages (Figure 13.1*a*), or as a stage contained within the developing egg. If there is a nauplius larva, the arthropod is a crustacean, but absence of the nauplius may be a secondary loss, especially in fresh water or on land.

Many characteristic crustacean features are associated with the aquatic habitat – for example, the hardening of the cuticle by calcium salts as well as by tanned proteins associated with the chitin – and are not found in land-living crustaceans such as woodlice. In aquatic crustaceans gaseous exchange for

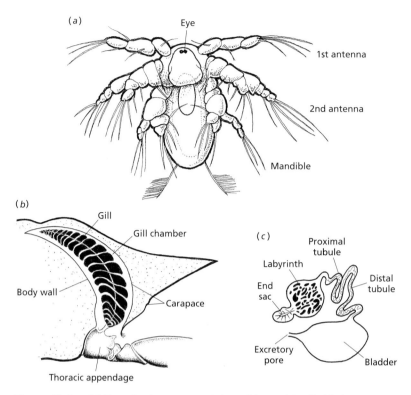

Figure 13.1. (a) Nauplius larva, ventral view; (b) a crab's gill; (c) the excretory organ (green gland) of *Astacus*, a crayfish.

respiration occurs over the whole body surface (in small animals) or over the gills, which are usually modified basal parts of appendages, covered in soft permeable cuticle (see Figure 13.1*b*). Oxygen is transported in the blood by the respiratory pigment haemocyanin. Excretion occurs partly by diffusion of ammonia over the gills or body wall and partly through mesodermal organs opening at the base of an anterior appendage (in crayfish these are called the 'green glands', see Figure 13.1*c*).

13.2 What are the main kinds of crustacean?

Among a great deal of convergent evolution, one clear evolutionary trend can be recognised, as follows:

Primitively, crustaceans were probably small filter-feeding 'shrimps' with a large number of similar appendages, and all the appendages behind the

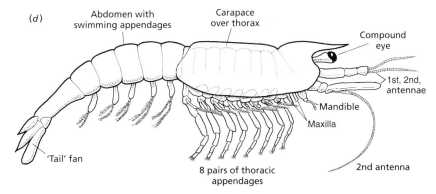

Figure 13.1 (cont.). (d) A generalised malacostracan (based on *Astacus*).

antennae took part in feeding, locomotion and respiratory exchange. From some such starting point, evolution has many times produced larger bottom-living forms that seek out and grasp their food in larger quantities. They have relatively shorter bodies and fewer, more specialised appendages. The head bears the segmental appendages shown in Figure 13.1*d*: two pairs of antennae, the mandibles (hard biting structures made from the base of the third head appendage) and two pairs of maxillae, with perhaps some anterior trunk appendages assisting feeding as 'maxillipedes' and perhaps also 'chelae' (claws). The anterior end or 'cephalothorax' becomes enclosed in a shell or 'carapace', a cuticle-lined chamber protecting the gills. The trunk appendages are specialised for walking or swimming. The extreme end-point of this trend is represented by crabs, where the often thick carapace covers a body so short that the abdomen is tucked underneath and the four pairs of walking legs are so close together that the crab can only walk sideways.

There are about 50 Orders of Crustacea. In this brief account, only a few of the great variety of forms can be introduced, mainly by annotated diagrams in Figures 13.2 to 13.4. Different lines of crustacean evolution such as colonisation of fresh water and land, parasitism and the variety of larval forms are then discussed.

The larger groups illustrated in Figures 13.2 and 13.3 are:

– **Branchiopods**, a very diverse group of small filter-feeding shrimps, many in fresh water, including the fairy shrimp *Chirocephalus*, (Figure 13.2*b*), the brine shrimp *Artemia* and the water flea *Daphnia* (Figure 13.2*c*).
– **Copepods** (Figure 13.3*c*), extremely common planktonic shrimps, mainly in the sea where they are a vital part of many food chains.
– **Ostracods** (Figure 13.3*d*), a very separate and ancient group of

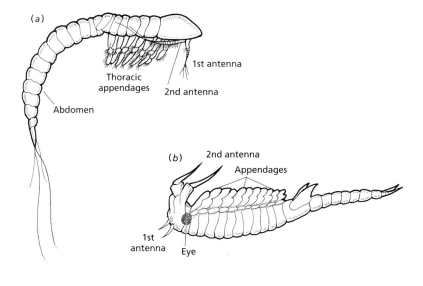

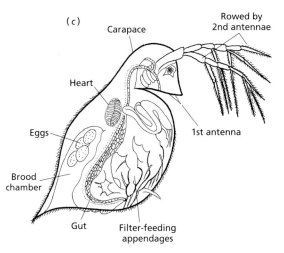

Figure 13.2. The diversity of Crustacea. (*a*) *Hutchinsonia macracantha*. Cephalocarida are probably the most primitive living crustaceans; less than 3 mm long, they live in marine mud. (*b*) *Chirocephalus grubei* (Branchiopoda, Anostraca), the freshwater fairy shrimp, swims with the dorsal side downward. Many leaf-like appendages, all used for feeding, swimming and respiration. On sale as pets, as 'sea monkeys'! (*c*) *Daphnia pulex* (Branchiopoda, Cladocera), the freshwater 'water flea'.

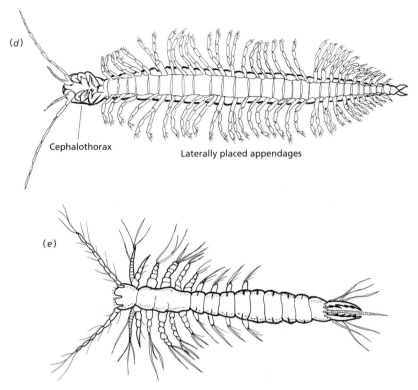

(d)

Cephalothorax

Laterally placed appendages

(e)

Figure 13.2 (cont.). (d) *Lasionectes entrichomas*. Remipedia are worm-like crustaceans less than 3 cm long, discovered in marine caves in 1981. They have uniform trunk segments with many similar appendages, but are thought to be secondarily simple. (e) *Derocheilocaris* sp., a mystacocaridan from the marine interstitial habitat. Only 10 species are known, less than 1 mm long.

crustaceans. Often less than a millimetre long, an ostracod is enclosed in a bivalve calcareous shell, even as a nauplius. The head bears unique appendages, the trunk has largely atrophied. Unlike water fleas, ostracods swim smoothly.

— **Cirripedes** are unique and very much modified crustaceans: barnacles are all sessile, fastened to a substratum by their heads and kicking their food into their mouths with their legs. They include the extremely common sessile barnacle of rocky shores, *Balanus* (Figure 13.3*e*) and the stalked ship's barnacle, *Lepas*. The carapace has become a thick mantle over the trunk, the head attaches to the substratum and when the barnacle is under water the upper shell plates open and the legs shoot out, casting a food-collecting net. They are cross-fertilised hermaphrodites. Like all

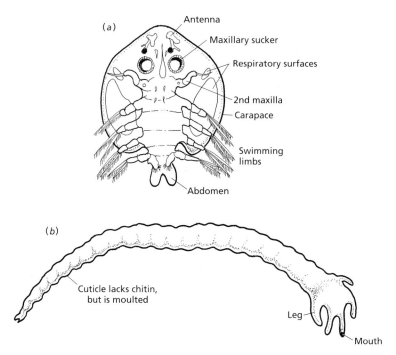

Figure 13.3. The diversity of Crustacea (continued). (a) *Argulus foliaceous* (Branchiura), ectoparasitic on fish: some 150 spp., less than 3 cm long. (b) *Cephalobaena tetrapoda* (Pentostoma) from a snake lung. There are ninety species of much modified crustacean parasites, once placed in a separate phylum, living in the nostrils or lungs of reptiles or other vertebrates.

sessile adult animals, they depend for their dispersal on motile larval stages.

Figure 13.4 shows some **Malacostraca**, a large group of Crustacea with some 23 000 species, not particularly 'advanced' but including the biggest and best-known species. Their anatomy is relatively constant: for example the body regions (head six, thorax eight, abdomen six segments) and the position of the genital openings (female sixth, male eighth abdominal segments).

Malacostraca include some primitive shrimps with a seventh abdominal segment (e.g. *Nebalia*), the mantis shrimps (e.g. *Squilla*), the Australian freshwater forms with no carapace (e.g. *Anaspides*) and two main groups:

– *Peracarida*, with a ventral brood pouch in which the eggs develop, including shrimps such as *Mysis* (Figure 13.4a), **amphipods**, laterally compressed, for example sandhoppers such as *Gammarus* (Figure 13.4b),

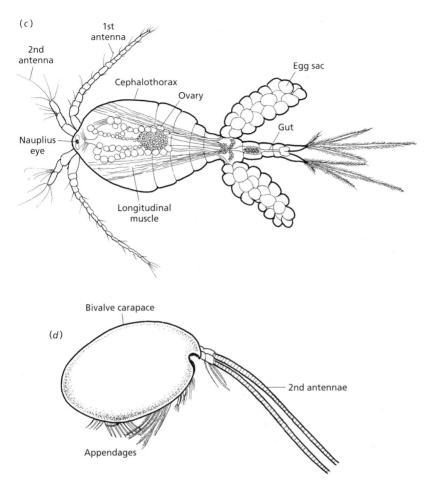

Figure 13.3 (cont.). (c) *Macrocyclops albidus* (Copepoda). There are some 8500 species and vast numbers of individuals of copepods in the marine plankton, forming an important basis for many food chains. They are less numerous in fresh water. (d) *Heterocypris* (Ostracoda). About 1 mm long, ostracods are enclosed in a bivalve carapace. They are rowed along by their antennae, in the sea or in freshwater.

isopods, dorsoventrally compressed, e.g. *Ligia* at the top of the shore and the woodlice (Figure 13.4*c*) on land.

– *Eucarida* with no brood pouch, the carapace being fused to the thoracic cuticle. These are **euphausids**, krill, vital to southern oceans food chains (Figure 13.4*d*), and **decapods**, where the appendages of the first three thoracic segments become maxillipeds assisting feeding, leaving five pairs of locomotor appendages ('10 legs'). See Figure 13.4*e–g*.

(e)

(i)

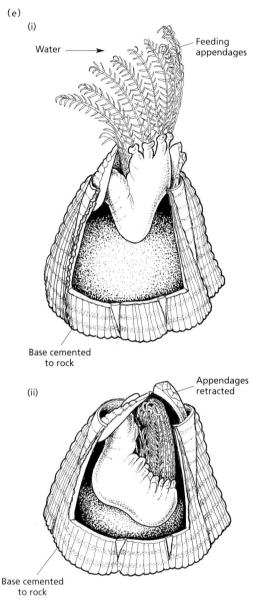

Water →

Feeding
appendages

Base cemented
to rock

(ii)

Appendages
retracted

Base cemented
to rock

Figure 13.3 (cont.). (e) *Balanus balanoides* (Cirripedia). Barnacles are unique, much modified sessile hermaphrodites, all marine. (i) At high tide, the appendages are shot out and kick food into the mouth. (ii) When exposed to the air, the appendages are retracted.

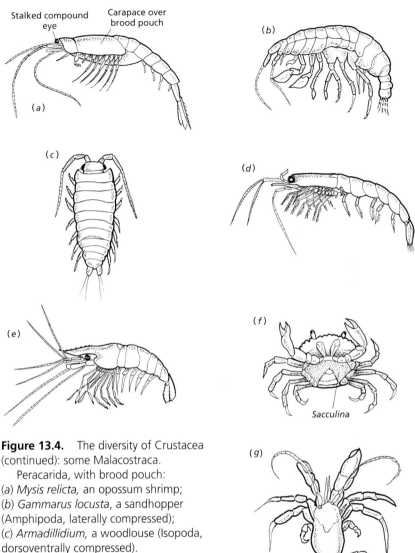

Figure 13.4. The diversity of Crustacea (continued): some Malacostraca.
 Peracarida, with brood pouch:
(a) *Mysis relicta,* an opossum shrimp;
(b) *Gammarus locusta,* a sandhopper (Amphipoda, laterally compressed);
(c) *Armadillidium,* a woodlouse (Isopoda, dorsoventrally compressed).
 Eucarida, with the carapace fused to the thoracic cuticle and no brood pouch:
(d) *Euphausia superba,* krill . Shrimp-like.
 Decapod eucaridans have five pairs of locomotor appendages as the first three thoracic pairs assist in feeding: (e) *Leander squilla,* a prawn; (f) *Carcinus maenas,* the shore crab, ventral view, with the rhizocephalan parasite *Sacculina carcini*;
(g) *Eupagurus bernhardus,* a hermit crab, showing the soft, reduced, asymmetrical abdomen.

13.3. How have crustaceans colonised fresh water and land?

13.3.1 Fresh water

Most branchiopods and the crayfish are freshwater animals, and most groups other than barnacles have a number of freshwater species, for example copepods, ostracods, amphipods, isopods, decapod shrimps and crabs. They provide an interesting range of examples of the mechanisms (described in Box 7.1) by which freshwater animals resist or counter the tendency for water to enter and ions to leave their bodies. Crustacea can readily make the cuticle almost impermeable to water except over the gills. The body fluids may be considerably more dilute than in sea water. Dissolved salts depress the freezing point in direct proportion to their concentration. The depression of the freezing point is 1.8 °C in sea water and also in the internal body fluids of most marine crustaceans, but 0.8 °C in the crayfish *Astacus* and only 0.3 °C in the water flea *Daphnia*, as a small animal with a relatively larger surface area needs to be as dilute as possible. Crustaceans able to penetrate estuaries (e.g. the shore crab *Carcinus*) can tolerate considerable internal dilution by liberating stored amino acids from muscle cells into the blood. In fully fresh water, however, energy-consuming active ion uptake occurs at the gills (or over the whole body surface). This replaces ions leaving by diffusion, and excess water is removed by the excretory organ. A few freshwater crustaceans are also able to reabsorb ions in the excretory tubule, for example the amphipod *Gammarus pulex* and the crayfish *Astacus*, which, unlike marine crustaceans, have a distal portion of the tubule (see Figure 13.1*c*).

Most freshwater crustaceans have direct development. Rather than hatch as small fragile larvae, they brood large yolky eggs that hatch at a later stage, and the eggs may be very resistant to desiccation, hatching when supplied with water. The Chinese mitten crab *Eriocheir sinensis* by contrast migrates up and down estuaries during the life cycle, returning to the sea to breed, having therefore to withstand great osmotic change. It has the most impermeable crustacean cuticle known. Osmoregulation occurs at the gills, whereby ions subject to active transport may either be taken up (when the medium is more dilute than the crab) or extruded (when the medium is more concentrated). *Eriocheir* at the same time is an osmoconformer in that it can tolerate great changes in body fluid concentration. When the blood is diluted, amino acids released from muscle cells are broken down to proline and then to ammonium ions (NH_4) at the gills, which are exchanged for incoming sodium ions (Na^+), saving energy. The urine is isoosmotic with the blood: it seems remarkable that there is no ion reabsorption in the excretory organ, but calculations show that hypoosmotic urine does not provide a significant energy saving in the fairly salty water where *Eriocheir* lives most of the time.

13.3.2 Land

Box 7.1 outlines the problems facing animals colonising land and the routes by which the transition was made. Few crustacean groups have fully terrestrial members. There are some land crabs, one family of amphipods and a profusion of isopods (there are at least 1000 species of woodlice). Habitats allowing semiterrestrial existence were probably crucial to the transition: present-day semiterrestrial crustaceans include the fiddler crabs (e.g. *Uca*), which spend much of their time in the air but retreat to water-filled burrows, and isopods such as *Ligia*, seeking out damp crevices high on rocky shores.

Terrestrial crustaceans have body fluids with high osmotic pressure, which at once suggests that they evolved not by way of fresh water but overland from the sea, perhaps over salt marsh or through mangrove swamps.

13.3.3 Land crabs

Land crabs are relatively large and impermeable invertebrates, not immediately liable to dry out. The stiffened gills do not entirely collapse in air and therefore allow some gaseous exchange, and part of the very vascular gill chamber functions as a lung. However, land crabs must return to the sea to breed, unlike amphipods and isopods that retain their eggs in a brood pouch where they develop directly.

13.3.4 Amphipods

Amphipods of the family Talitridae are common above the strand line on sandy beaches: sand hoppers live in burrows above high tide mark by day and emerge at night when they feed on detritus. There is a series of species with progressively more enclosed respiratory organs, but terrestrial colonisation seems to depend mainly on tolerance of variation in body fluid composition and on behavioural adaptations.

13.3.5 Isopods

Isopod gills are ventral flat plates at the posterior end of the body, covered and enclosed by the last appendages (uropods). Air-filled 'pseudotracheae' are probably crucial for breathing air with minimal water loss. Woodlice tolerate changes in internal concentration, and have the unusual ability to excrete by puffing out gaseous ammonia. Woodlice have become very successful terrestrial animals, but cannot live in hot dry air. They have to have behavioural adaptations that keep them in damp places.

13.4 What are the special features of parasitic crustaceans?

Parasitism has arisen many times separately in Crustacea. Ectoparasites, living mainly on the gills or the outer surface of fish, are recognisable Crustacea with certain appendages modified for attachment to the host (e.g. Branchiura, several separate Orders of copepods and also Tentaculocarida, which are a whole newly discovered Class of deep sea crustaceans). Endoparasitism demands radical reorganisation, since an unmodified exoskeleton would be a barrier between parasite and host. For example, the 90 species of Pentastoma (Figure 13.3*b*), parasites in the nostrils or lungs of reptiles, were formerly placed in a separate phylum. The cuticle lacks chitin but is moulted) and there are no respiratory, circulatory or excretory organs. *Sacculina* is an even more modified example. The body consists of an external sac of eggs under a crab abdomen (see Figure 13.4*f*) and a series of food-absorbing tubes inside the crab. It is revealed as a crustacean only by the nauplius larva, which has the 'horns' characteristic of cirripede nauplii (see Figure 13.5*a*). The host is not killed, but the male is progressively more feminised at each moult as the parasite consumes the androgenic gland, the source of male hormones. Other Rhizocephala are parasites with little more than fungus-like rooting and food-absorbing systems inside the host's body. Metamorphosis in these animals is perhaps more drastic than in any other parasite, with no larval tissue being carried through to the adult.

Copepod endoparasites tend similarly to be very much reduced. Among them Monstrilloida, in a variety of invertebrate hosts, are unusual in being parasitic only as juveniles. The adults are free-living and recognisable as crustaceans, but lack mouthparts. Yet other crustaceans have a division of labour between the sexes, only the female being parasitic.

In general, Crustacea require great specialisation (in the guise of 'degeneration', see Box 6.1) to become successful endoparasites. Success is facilitated by the need to moult repeatedly during the life history, since a series of larvae can effect transfer between hosts.

13.5 What is the role of crustacean larvae?

The need to moult provides opportunities for changes in form, often striking enough to be called metamorphosis. Different forms in the life cycle between the egg and the sexually mature adult allow a division of labour: most typically crustacean larvae are feeding stages in a habitat different from that of the adult, avoiding competition with the adult, and serving also to disperse the animals. Most marine Crustacea hatch as planktonic larvae, in the surface waters, which are rich in food and have currents separate from those lower in the sea. Freshwater Crustacea seldom have larvae and terrestrial ones never do (see above). Large yolky eggs contain and feed the developing young, often within a brood pouch.

13.5.1 Examples of larvae and life cycles (see Figure 13.5)

The nauplius is most commonly the first larval stage. It has three pairs of appendages, the smallest number that can filter-feed. Even within a directly developing egg, a pause at a recognisable nauplius stage can often be detected.

- **Copepods:** nauplius: metanauplius (four pairs of appendages): a series of larvae progressively more like the adult. Metamorphosis is very gradual, with adults remaining in the larval habitat. They retain some larval characters, for example there is no carapace and only a simple 'nauplius' eye.
- **Cirripedes** have a far-reaching metamorphosis. After the nauplius (distinctive with 'horns', (Figure 13.5a) and metanauplius there is a 'cypris' larva (Figure 13.5b) that does not feed but selects the adult habitat, seeking a rock where a protein 'message' has been left by an earlier barnacle occupant. In a sessile animal, larvae are particularly important for dispersal and habitat selection.
- **Malacostraca** nearly all hatch beyond the nauplius stage. Freshwater malacostracans have direct development, marine ones often hatch as a

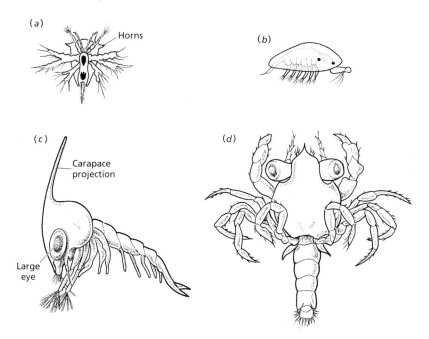

Figure 13.5. Some crustacean larvae: (a) cirripede (barnacle) nauplius; (b) cirripede cypris, which searches for a site and settles; (c) crab zoea, the first larval stage; (d) crab megalopa, a later stage, with abdomen still extended

zoea (Figure 13.5c) followed by a larva more like the adult, for example the crab zoea is followed by a megalopa (see Figure 13.5d), a small crab with the abdomen extended instead of being tucked underneath.

Crustacea, in summary, are a very large and diverse group of essentially aquatic arthropods. The arthropod groups to be considered next have members far more thoroughly adapted to life on land.

14

Chelicerata and Myriapoda

While Crustacea are aquatic arthropods with a few terrestrial members, Chelicerata are terrestrial arthropods with a few aquatic representatives. They demonstrate one version of the adaptations necessary for a terrestrial arthropod.

14.1 What are chelicerates?

They are arthropods whose segmental plan is different from others in that they have no head, only one pair of mouthparts (yet they are predators) and, unlike all other groups of arthropods, no antennae. The body is divided into an anterior prosoma and a posterior opisthosoma. The prosoma usually consists of eight segments. The first segment never bears appendages, the second bears paired chelicerae (feeding and grasping structures) and the third bears paired pedipalpi, (sensory, prehensile or reproductive appendages). The fourth to seventh segments bear walking legs (i.e. four pairs), which have small pincers (chelae) at their ends.

Chelicerates comprise (see Figures 14.1 and 14.2):

- **Merostomata**, aquatic forms with five or six pairs of opisthosomal appendages modified as gills, consisting of:

  - **Euryptera** (sea scorpions), all extinct.
  - **Xiphosura** (represented today by the horseshoe crab *Limulus*).

- **Pycnogonida** ('sea spiders'), which only superficially resemble spiders.

- **Arachnida**, all the terrestrial chelicerates, including:

  - Scorpions
  - Pseudoscorpions
  - Solifugae (sun spiders)
  - Opiliones (harvestmen)
  - Acari (mites and ticks)
  - Araneae (spiders)

14.2 Why is *Limulus* of special interest?

Comparisons can be made both with marine crustaceans and with the terrestrial chelicerates. *Limulus* (Figure 14.1*a,b*) is one of the three genera of marine chelicerates alive today. It burrows into the upper layers of sand in shallow seas, using the shield-shaped prosoma that protects the ventral appendages: these are the chelicerae and five pairs of walking legs (the front pair corresponds to the pedipalps of other chelicerates). At the posterior end of the gill-bearing opisthosoma there is a long spike, used for righting the animal and pushing.

14.2.1 Feeding

Molluscs, worms or other animals are picked up by the chelate appendages and held against a ventral groove in the body. The food is chewed not by mandibles but by hardened 'gnathobases' of the posterior walking legs and passed forward to the mouth, where it is swallowed and then further crushed in a gizzard.

14.2.2 Respiration

Opisthosomal appendages are modified into 'gill books' (Figure 14.1*b*) that move to and fro, making a continual stream of water over the gills. The blood contains the respiratory pigment haemocyanin, which has a low oxygen affinity, able to function at extremes of environmental oxygen pressure. Blood enters the long heart through eight pairs of ostia with valves. The further circulation is complex, with many arteries opening into the haemocoel.

14.2.3 Excretion and osmoregulation

As in Crustacea and other aquatic arthropods nitrogenous waste is removed partly as ammonia diffusing over the gills and partly through mesodermal glands at the base of a pair of appendages (here 'coxal glands' beside the back legs). *Limulus* can tolerate a wide range of salinities: the gills can take up ions when the medium is diluted and the coxal glands can produce urine hypoosmotic to the blood. The assumption, based on the similarity of the cellular structure to that of crustaceans, is that the cortex forms an ultrafiltrate that is modified in the medulla and end sac.

14.2.4 Reproduction

The sexes are separate. Males and females aggregate in shallow water at high tide when the moon is full. Fertilisation is external: the male mounts the female and pours sperm over the thousands of eggs as she lays them. The eggs develop in the sand and hatch as 'trilobite' larvae that swim and burrow and, in a series of moults, undergo gradual metamorphosis.

14.2.5 The compound eyes

First studied early in the twentieth century, the relatively uncomplicated compound eyes of *Limulus* were the primary source of information about these arthropod structures, and remain the context in which reference to *Limulus* is most common. *Limulus'* compound eyes do not need to work very fast. They are used not to find food, nor to escape from predators, but to detect the movement of other horseshoe crabs across the field of vision; they are apparently used only to achieve the aggregation of mating males and females in shallow water at high tide. Accordingly, the structure and performance of this eye (Figure 14.3*a*) is relatively simple and accessible to study.

In summary: *Limulus* is a specialised modern animal, but none the less it gives us an aquatic 'base-line' for considering the Arachnida.

14.3 What are pycnogonids?

Pycnogonids are small (1–10 mm long) arthropods crawling on sponges, hydroids or bryozoans, or deep sea forms up to 40 mm long; about 1000

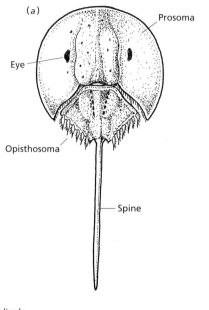

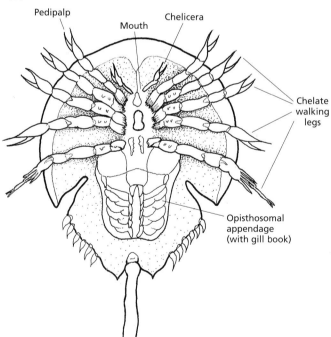

Figure 14.1. Aquatic Chelicerata: (a) *Limulus polyphemus*, the horseshoe crab, dorsal view; (b) *Limulus* in ventral view, showing appendages. Opisthosomal appendages bear gills, in a respiratory chamber.

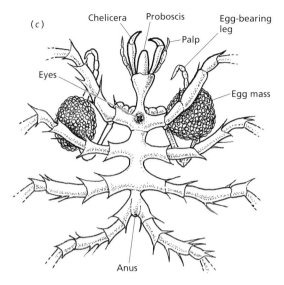

Figure 14.1 (*cont.*).   (*c*) *Nymphon rubrum*, a pycnogonid.

species are known. They are called sea spiders because they typically have four pairs of very long legs, each borne on a large extension of the supporting segment (Figure 14.1*c*). Usually they have little else: the body typically consists only of a narrow prosoma with an anterior sucking proboscis (they are carnivores or detritus feeders). The anterior appendages are sufficiently like chelicerae and palps to diagnose them as chelicerates, but they are highly modified and of unknown relationship to Merostomata and arachnids. The opisthosoma is remarkably reduced or absent, the digestive organs and the gonads extend into the legs and there are no special respiratory or excretory structures. The ova ripen in the legs of females and the eggs are carried by the males on modified appendages in front of the first pair of walking legs.

14.4 What are arachnids?

All terrestrial chelicerates are included and the group is thought to be monophyletic. Ninety-five per cent of the species are spiders, mites and harvestmen. Typically arachnids are predators feeding on other arthropods. Rather than catching them by sustained rapid locomotion, they use poisons – neurotoxins that paralyse their prey. The poison may come from the chelicerae or pedipalps or from a posterior sting, i.e. arachnids may be dangerous at both ends.

Digestion often occurs outside the body, with the posterior end of the oesophagus modified as a 'pumping stomach' to suck in the digested food.

14.4.1 Scorpions (1000 spp.)

Scorpions are large arachnids, up to 9 cm (one species, 21 cm) long with pedipalps bearing large characteristic chelae and with an extended segmented opisthosoma ending in a sting (Figure 14.2*a*). They are nocturnal hunters, burrowing or hiding under damp logs or in leaf litter. Most require humid conditions but some can live in deserts. They have a number of characters primitive among chelicerates, such as compound eyes and a nerve cord with separate segmental ganglia rather than a fused ganglionic mass. They resemble the fossil aquatic eurypterids ('sea scorpions') and are known from the Silurian as fossil aquatic scorpions with gill books protected by opisthosomal plates. Modern scorpions are all terrestrial, with the gill books replaced by lung books and stout sensory setae replaced by long thin sensory hairs. Scorpions have unique paired sensory combs ('pectines') that may be chemosensory or may detect vibrations in the ground.

The main characteristics of **pseudoscorpions, sunspiders, harvestmen** and **mites** and **ticks** are indicated in Figure 14.2.

14.4.2 Spiders (32 000 spp.)

Spiders are mostly 2–10 mm long, but tarantulas measure up to 90 mm. They are all carnivorous, feeding almost exclusively on insects. They are extremely successful terrestrial predators, largely on account of their production and use of silk (see below). The prosoma is convex dorsally, with marginal eyes (typically eight). The chelicerae seize the prey and paralyse it with poison, the pedipalps bear projections that manipulate food and may wrap it in silk (until it is ready for external digestion by enzymes poured out by the spider) and there are four pairs of walking legs (Figure 14.2*f*). A characteristic waist or 'pedicel' separates the prosoma from the opisthosoma, where soft parts of the body are housed and silk is manufactured and released.

14.5 How did arachnids colonise the land?

Land was probably colonised many times separately; the relationship between arachnid Orders is entirely obscure. Scorpions in the Silurian are the earliest

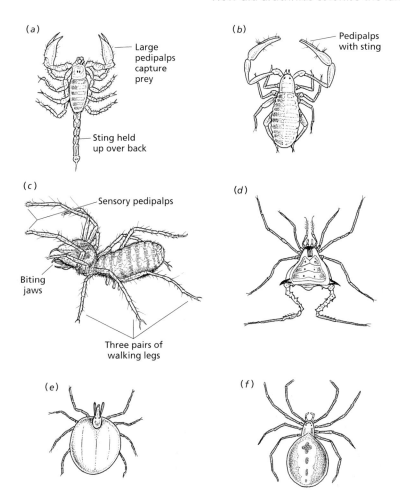

Figure 14.2. Arachnida: (*a*) A scorpion. (*b*) A pseudoscorpion: there are about 2500 species, less than 8 mm long, in crevices and leaf litter. Their resemblance to scorpions is only superficial. (*c*) A solifugid: about 900 species, about 7 cm long. They are fast running desert predators, with long sensory pedipalps held out in front and powerful biting chelicerae but no poison. (*d*) An opilionid: there are 4500 species of 'harvestmen' 5–20 mm long, predators or scavengers in damp grass or forests. (*e*) A tick: there are 30 000 species of mites, ticks and other Acari, very numerous as individuals, less than 1 mm long. They are free-living or ectoparasitic. (*f*) A spider.

known terrestrial arachnids; spiders are found in the early Carboniferous. The blood of all arachnids has a high osmotic pressure and a composition similar to that of *Limulus*, with sodium and chloride ions predominating: this suggests marine ancestry and direct colonisation of land over the upper shore (see Box 7.1). Spiders are especially successful terrestrial arthropods. Insects had appeared in the Mid Devonian, and the relationship between the two groups must have been established early, with increase in spider numbers following the radiation of insects. Today spiders are major invertebrate predators, numerous in all terrestrial ecosystems; for example, a study of untreated grassland in late summer in Sussex, England, gave an estimate of 2.25 million spiders per acre (about 5.55 million per hectare).

Arachnids are adapted to terrestrial life mainly by physiological changes and also by their behaviour: relatively large arachnids often burrow, and smaller ones are 'cryptozoic', hiding in crevices or under rotting logs. Physiological changes (as in other arthropods) primarily concern methods of resisting desiccation by modifications to the cuticle, the respiratory system and the excretory organs; sense organs and reproductive methods also need to be different in a terrestrial habitat.

14.5.1 The cuticle

Like that of insects, the arachnid cuticle has an outer layer containing wax, the epicuticle, which to varying extents resists water loss: some scorpions and ticks are as impermeable as insects.

14.5.2 Respiration and blood

Ancestral gill books have been replaced by 'lung books' (also called ' book lungs') where perhaps as many as 750 leaflets full of blood are separated by air spaces (Figure 14.3*b*). Many arachnids have also developed tracheal systems, resembling those of insects, except that in arachnids such systems convey oxygen not to every cell but to the blood, where it is carried round the body by the respiratory pigment haemocyanin. Spiders may have various forms of tracheae as well as lung books; some have tracheae alone. The tracheae open by spiracles, which can be closed to restrict respiratory water loss. The most highly developed tubular tracheae occur in the fast-running solifugids, where air in the tubules is moved by a degree of muscular ventilation.

The arachnid heart lies dorsally in the opisthosoma (Figure 12.4*a*). Primitively blood enters it by many ostia, but where tracheae take over from lung books the heart becomes simpler; it has only two pairs of ostia in small tracheate spiders. The heart pumps the blood forward by way of a series of dorsal arteries to the haemocoel of the prosoma, where increase in blood pressure protrudes the walking legs: the blood pressure may double when jumping

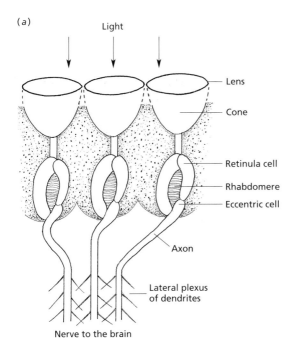

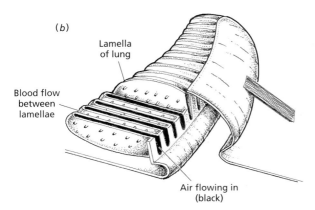

Figure 14.3. (a) Three ommatidia of the compound eye of *Limulus*; (b) the book lung of a spider.

spiders jump. The constricting waist of spiders allows the pressure in the prosoma to be independent of that in the opisthosoma, and in solifugids the two circulations can be separated by closure of a diaphragm. That the blood serves both to transport oxygen and to extend the legs can limit the activity of arachnids: if forced to sustain activity they become short of oxygen.

14.5.3 Excretory organs

The coxal glands serve for osmotic regulation. Study of a tick reveals ultrafiltration into a sac, sucked by external muscles. Nitrogen excretion occurs not in the coxal glands but in 'Malpighian tubules', analogous to those of insects (see Chapter 15), or in the walls of the midgut. The nitrogenous end-point is guanine, a purine even more insoluble than uric acid.

14.5.4 Sense organs

In arachnids, sense organs are of three kinds: hairs on the cuticle, modified to detect mechanical or chemical stimuli; eyes, usually simple rather than compound and often spread over the body; and slit sensory organs, pits in the cuticle each covered by a thin membrane bulging inward towards a sensory hair, able to detect slight changes in the tension of the cuticle. These last are used largely as proprioceptors, i.e. to inform the animal about the degree of contraction of its own muscles.

14.5.5 Reproduction

Fertilisation is internal, by direct transfer in harvestmen but more usually by way of a package of spermatozoa done up in silk. These packages are deposited on the ground by scorpions, pseudoscorpions and some mites, but spiders (uniquely) handle the package with their pedipalps and, after ritualised courtship, insert it into the female. Courtship allows the male to avoid being mistaken for prey: in some species he risks being eaten by the female after mating but, contrary to popular belief, this is rare. Development is direct, in large eggs with the yolk central and uncleaved, as in insects.

14.5.6 Silk

Silk is a protein, fibroin, containing an unusually high proportion of the amino acids glycine and alanine. It is extremely strong: 10 times stronger than collagen, as strong as nylon but more elastic. The silk glands include a tubule in which the raw material of silk is made more acid and water is withdrawn to improve gel formation. The glands are at the posterior end of the opisthosoma, opening by mobile spinnerets; these and the mobility allowed by the narrow waist enable the spider to place the silk, which is not squirted into the air but attached to a surface. The silk is then hardened because the weight of the spider pulls out the thread, which alters its molecular configuration. This process can be checked by reduction of pressure in the opisthosoma.

Spiders use silk for sperm transfer, egg cases, nest linings, draglines and for dispersal: 'ballooning' spiders can drift over long distances. Probably it evolved initially as a solution to reproductive requirements, but most characteristically it is used for food capture. The 'orb' web of the garden spider, characteristic of the large family Araneidae, is made of two kinds of silk, one constituting the framework and the radial threads, between which is wound a spiral of the other kind, 'viscid' silk bearing glue droplets. Solid connections form wherever the threads cross. The viscid silk is extensible enough to catch quite large insects (not too large, they might be fierce). The spider sits at the centre or runs over the web (it does not stick to its own web partly because the legs secrete an oily substance and mainly because the principal 'spokes', along which the spider runs, are not sticky). Web spiders use their eyes very little. Their mechanoreceptors are extremely sensitive to disturbance of the web, and their slit sense organs are grouped as 'lyriform organs' that enable spiders to judge both the angle of change in tension in the cuticle and the distance they have covered over the web.

Different spiders make differently shaped webs, not necessarily two-dimensional: funnels, triangles, domed, sheets, placed horizontally or vertically. An elaborate web can be constructed within an hour, often only after the old web has been eaten. Spiders given caffeine or amphetamines make odd webs, deformed in ways constant for the particular drug. Young spiderlings make perfect webs without having to learn, yet there is some plasticity: remarkably, despite the role of gravity in normal web construction, spiders in space capsules can make perfect webs. Cursorial spiders (wolf and crab spiders, etc.) and jumping spiders have highly developed eyes and hunt by sight, often as ambush predators. They produce a dragline of silk, and may use it as a snare.

In summary, the production and use of silk is the key to the great success of spiders as predatory terrestrial animals.

MYRIAPODA

14.6 What are myriapods and how do they move?

Behind the head, a myriapod consists of a series of similar segments, each bearing one or two pairs of walking legs, with no division into regions such as thorax and abdomen, prosoma and opisthosoma (Figure 14.4). The head is

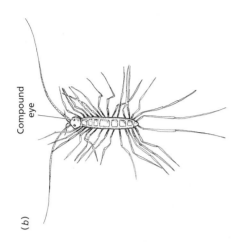

(a)

Antenna

Poison claw

Leg in
recovery
stroke

Leg in
effective
stroke

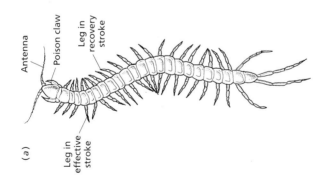

(b)

Compound
eye

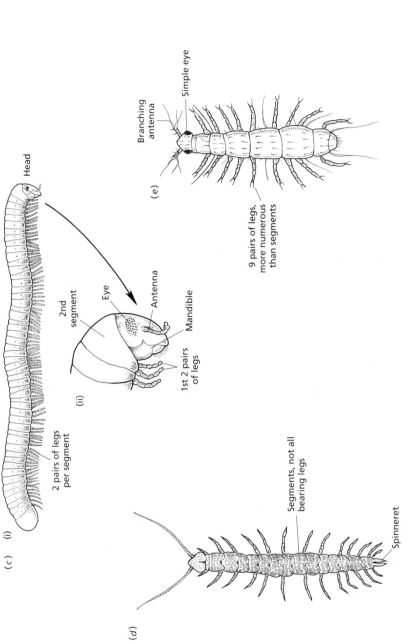

Figure 14.4. Myriapoda.

Chilopoda, centipedes: (a) a scolopendromorph, showing locomotor waves; (b) *Scutigera*, showing legs progressively longer posteriorly.

Diplopoda, millipedes: (c) (i) *Julus* and (ii) head of *Julus* enlarged.

(d) Symphyla are detritus feeders in the soil, up to 8 mm long, superficially like centipedes.

(e) Pauropoda resemble very small millipedes. Less than 2 mm long, they are detritus feeders in damp leaf litter or soil.

distinctive: behind the first segment without any appendages and the second segment with a single pair of antennae, the mouthparts are modified to form an upper lip, mandibles and a lower lip (labium) within which a buccal cavity is enclosed. Except in one group of centipedes, there are no compound eyes. All known myriapods are terrestrial.

14.6.1 Chilopoda

Centipedes or Chilopoda are predatory carnivores with the first pair of walking legs forming poison claws. About 3000 species are known, and they are up to 270 mm in length. Their soft bodies are flattened dorsoventrally with the legs projecting laterally; they can insert their flat bodies into crevices and spaces in litter. 'A hundred legs' is not an accurate description: there may be many fewer or as many as 177 pairs, all alike. Some run very fast, with a metachronal rhythm similar to that of polychaetes (Figure 14.4a and Chapter 9), with the wave passing from back to front. Tripping over the legs is a hazard: long and short legs may alternate, or more posterior legs may be longer and overlap (*Scutigera*, Figure 14.4b). The fast-moving scutigerids are the only myriapods to have compound eyes and respiratory pigments.

14.6.2 Diplopoda

Millipedes or Diplopoda feed on plants or decaying material. Their bodies are usually round in cross-section with a hard cuticle (Figure 14.4c). The body is held straight and the legs are ventral; unlike the fast-running centipedes, millipedes characteristically walk slowly, with a powerful thrust. They can push themselves into litter and decaying logs and most are able to bend dorsoventrally and curl their bodies into spirals. About 10000 species are known, up to 280 mm in length, living in dark humid places. They do not have a thousand legs, but each apparent segment is double and bears two pairs of legs

14.6.3 Symphyla and Pauropoda

The main characteristics of Symphyla and Pauropoda are indicated in Figure 14.4d and e. The interrelationships of these groups is controversial, except that the symphylans are close to centipedes and pauropods to millipedes. Myriapods as a whole are not now thought to be monophyletic, nor to be particularly closely related to insects (see Chapter 20).

14.7 How well are myriapods adapted to life on land?

How myriapods invaded the land is not known; possibly by the interstitial route between sand grains. The earliest fossils found, as late as the Silurian, are in terrestrial deposits and there is nothing to indicate a freshwater origin. The cuticle is not very impermeable compared with that of arachnids, often containing little wax. Excretion is by Malpighian tubules, less highly developed than those of insects. Respiration is by tracheae; unlike insect tracheal systems they do not convey oxygen to each cell but to the blood, as in arachnids. The pattern of tracheal systems is related to the level of activity: millipedes have spiracles on all segments behind the third and short tufts of tracheae, while centipedes, more active, have fewer spiracles but a pair of lateral tracheal trunks giving off branching tubes, as in insects. The spiracles cannot be closed and are therefore an important limit to water conservation.

In summary: myriapods are only incompletely terrestrial, with most species confined to damp habitats. For arthropods even better adapted to land life than the spiders, one must look to the insects.

15

Insecta

Most animals are insects and one group, the beetles, is the largest known Order of animals. There are vast numbers of individuals, about a million described species and many times that number awaiting description. They are by all criteria outstandingly successful, primarily on land, where the close relationship between insects and flowering plants first evolved, but many species occur in fresh water, often as immature stages. Unlike crustaceans, there are few species in the sea. The following account first explains that insects owe their success as terrestrial arthropods to a number of special characteristics, such as their water conservation mechanisms (already introduced in the previous three chapters) and the power of flight, which is so important that it demands a relatively full discussion. The life cycle is introduced to show that it provides flexibility, especially in those insects where there is a pupa, a transitional form that may provide a resting stage additional to the egg. An indication of the range of insect Orders is given, the evolution of social insects is briefly discussed and finally the contribution of the fruitfly *Drosophila* to our knowledge of Genetics is mentioned.

15.1 What is an insect?

Insects typically have three pairs of legs, and accordingly are called 'hexapods'. The adult body is divided into head, thorax and abdomen. The head bears a pair of compound eyes. The paired segmental appendages are one pair of antennae (second segment) and typically mandibles (fourth), maxillae (fifth) and fused second maxillae (the 'labium') on the final, sixth, head segment. The form of the mouthparts varies greatly. Each of the three thoracic segments bears a pair of legs, and typically there is a pair of wings on each of the two posterior segments . The abdomen has 9 to 11 segments and no appendages (except genitalia or their derivatives). It bears the gonopores and the gut opens at the posterior tip. Excretion, including water regulation, is carried out by Malpighian tubules at the junction between the mid and hind gut. Respiratory gases are transported between the outside air and every cell of the body by tracheae, branching tubes lined with cuticle that open at the body surface by holes called spiracles. The small size of insects (the largest present-day species approach the size of a mouse and most are very much smaller) is dictated partly by the requirements of the tracheal system but probably even more by the difficulty of supporting the body in air when the exoskeleton is shed at moulting. Insects usually dominate ecological niches requiring small size.

15.2 Why are insects such successful land animals?

Small size is a crucial attribute of insects. Competition with vertebrates can be avoided by animals two orders of magnitude smaller: ecological niches also are related to size, and insects have been able to dominate those that require small-sized occupants. A further advantage of small size is rapid reproductive turnover, which not only accelerates population increase but also allows faster evolutionary change. The outstanding physiological advantage of being small is that the tracheal system can supply oxygen directly to every cell of the body (Figure 15.1*a*). Air enters the tracheae through thoracic and abdominal spiracles and diffuses (aided in the larger insects by ventilation movements) along the much branched and anastomosing tracheal tubes to the thin-walled 'tracheoles' which penetrate between or even into cells (Figure 15.1*b*). Unlike arachnids, insects rely totally on the tracheal system for transport of oxygen: the blood plays no part in respiration and contains no respiratory pigment. (As always with insects, generalisation is rash: there are some midge larvae, 'blood worms', containing haemoglobin.) The tracheae are lined by cuticle, with spiral thickening that keeps the tubes open. At moulting all tubes except the tracheoles have to be shed completely and replaced.

Small size, however, has the overwhelming disadvantage that the surface area of the animal is very large in proportion to the volume, so that on land there is great risk of desiccation . The success of insects depends upon water conservation, which is achieved to an unprecedented extent.

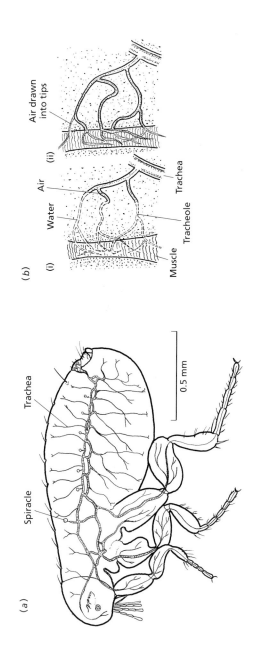

(a)

Trachea

Spiracle

0.5 mm

(b)

(i)

Water Air

Trachea

Tracheole

Muscle

(ii) Air drawn into tips

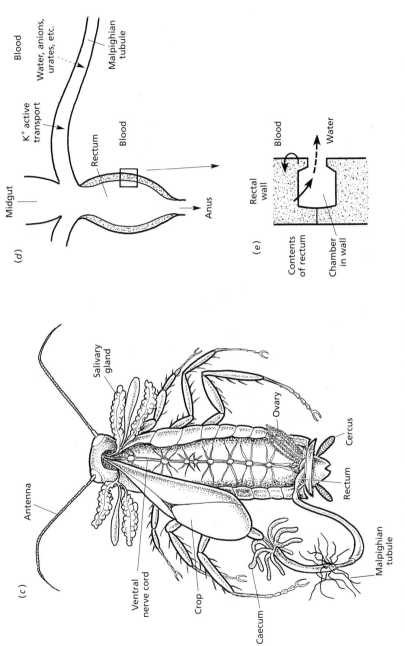

Figure 15.1. Insect tracheal systems and Malpighian tubules. (*a*) One half of the tracheal system of the flea, *Xenopsilla*. (*b*) Tracheoles supplying a muscle fibre (i) at rest and (ii) when active. (*c*) The gut of a cockroach showing the Malpighian tubules. (*d*) Diagram of a Malpighian tubule in relation to the blood and the rectum. (*e*) Diagram showing ion and water movements within the rectal wall. Solid arrows denote active ion transport; dotted arrows in (*d*) and (*e*) denote passive movements of ions or water.

15.2.1 The cuticle

The cuticle (see fuller account in Chapter 12) is almost completely waterproof, due to the organisation of the wax layer. An ordered molecular structure is suggested by the effect of temperature change on the permeability of the epicuticle, but the details are still controversial.

15.2.2 The tracheal system

No other animals can take oxygen from the air with so little accompanying water loss. Only in the tracheoles are gases exchanged over a damp surface where water could evaporate, and the tracheoles are far removed from the body surface. The distant spiracles are the only sites of water loss from the body, and they are kept closed by valves for much of the time.

15.2.3 The Malpighian tubules and the rectum

These organs are uniquely efficient at regulating water loss during excretion of metabolic waste in the very insoluble form of uric acid. Malpighian tubules are long thin tubes arising at the junction of the mid and hind gut and extending into the haemocoel, where they are bathed in blood (Figure 15.1c).

Most insects have no appreciable blood pressure. Therefore waste cannot enter the excretory tubule by ultrafiltration, as occurs in most other animals, including most groups of arthropods. Instead, ions are actively transported from the blood into the upper part of the tubule; often only one kind of ion is moved actively and ions of the opposite charge follow passively (Figure 15.1d). Potassium is frequently the ion moved actively, chloride follows, with many other anions including urates (the salts of uric acid), amino acids, sugars and other metabolically useful substances. The resulting osmotic potential draws water also into the tubule and the solution passes down to its base where it enters the rectum. Here the pH is lower, which causes uric acid to crystallise out . The rectal walls are very muscular and are protected by cuticle from toxins and from abrasion. Some part of the wall is thickened to form compartments into which ions are actively transported (Figure 15.1e). Water from the contents of the rectum is then also drawn into those compartments and passes on through open channels into the blood. Metabolically useful substances are similarly reabsorbed. The rectum then squeezes out undigested food and uric acid crystals through the anus, with hardly any accompanying water. Some insects even absorb water from the unsaturated atmosphere through the rectum.

Malpighian tubules are under hormonal control, and their rate of ion secretion can be altered if water needs to be removed rather than conserved. For example, when the bloodsucking bug *Rhodnius* has had a large meal, the resulting stretching of the abdomen stimulates nerve endings in its wall, and these nerves cause cells in the middle thoracic ganglion to release a diuretic hormone into the blood. This can increase the rate of excretion a thousand-fold and water is lost at the anus until equilibrium is restored.

In summary: the tracheal system, the uniquely thorough removal of water from faeces and excretory material and above all the water-proofed cuticle enable insects to be outstandingly successful terrestrial animals.

15.3 How are insects able to fly?

Insects are the smallest flying animals. The essential requirements for flight are the same as those for birds: they need wings providing **lift**, an upward force opposing gravity, and **thrust**, a forward force opposing the backward drag which inevitably accompanies lift. While the essential requirements are the same, the ways in which flying animals obtain lift and thrust depend greatly upon their size.

15.3.1 Insect wings

The wings are paired extensions of the cuticle of the dorsal part of the second and third thoracic segments. These flaps of cuticle contain haemocoel and tracheae (Figure 15.2*a*). To expand the wings initially (at the final moult) blood is pumped into the haemocoel: it then dries out and nearly all the epidermal cells die, leaving the wings very light and strong.

15.3.2 Lift

Lift is provided in any wing, and accompanied by thrust, when the air moves faster over the upper surface than over the lower, provided also that the 'angle of attack' is positive (Figure 15.2*b,c*). A bird's wing in cross-section has the

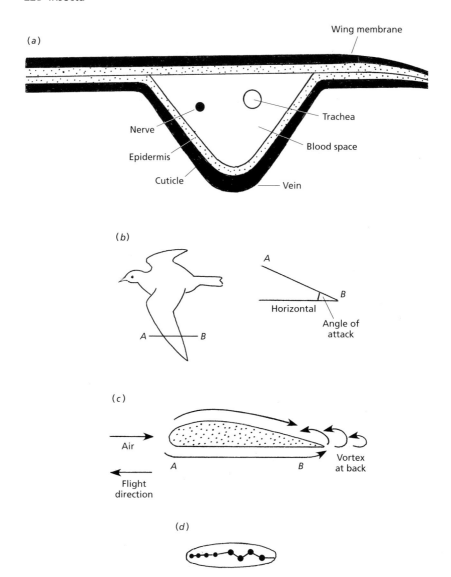

Figure 15.2. Insect wings and flight. (a) Transverse section (T.S.) of insect wing; (b) and (c) diagrams of a wing to show how an aerofoil may obtain lift; (d) T.S. insect wing with boundary layer; (e) (i) clap, (ii) fling, and (iii) end of fling, showing air movements; (f) the leading edge vortex.

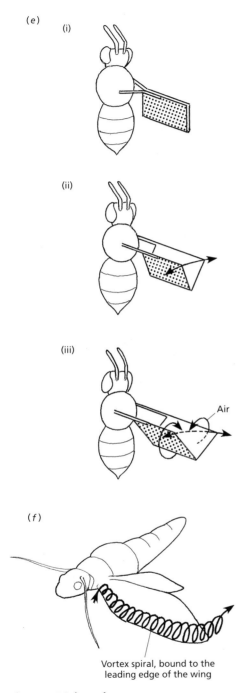

(e)

(i)

(ii)

(iii)

Air

(f)

Vortex spiral, bound to the
leading edge of the wing

Figure 15.2 (cont.)

shape known as an 'aerofoil section', the curved upper surface providing a longer path for the air flowing over it. The airstream over the lower surface therefore reaches the back of the wing first, and air is sucked upwards to meet the airstream over the upper surface. A vortex is set up at the back (the 'trailing edge'), pulling back the air over the upper surface so that it moves faster, providing lift. A large bird can glide, since there is a great difference between the rates of air flow over its upper and lower wing surfaces. In large flying animals, flapping is not essential to obtain lift.

The small size of insects demands a rather different mechanism, as detailed below.

- **Flapping**: an insect, like a small bird, must flap to obtain sufficient lift, by increasing the rate of air flow over wings moving much faster than the main body (lift is proportional to the square of the velocity). The small size of insects is favourable for flapping, because the power : weight ratio is high.
- **The boundary layer** of inert air sticking to the wing surfaces constitutes the greatest problem for a very small flier. Owing to the viscosity of air any wing has a boundary layer, of the same width for any size of wing: a boundary layer of 1 mm is negligible in birds, but relatively enormous in insects. In consequence of this large 'viscous drag', the lift : drag ratio (on which the speed and efficiency of flight depends) is at best only about one-tenth of that in birds. Any increase in surface area increases viscous drag, and accordingly insect wings are flat plates, not aerofoil sections. However, an aerofoil section is produced when a flat plate with an adhering boundary layer is moved through the air, and this shape enables lift to be produced.
- **Obtaining sufficient lift** remains difficult for a very small flier, as there is very little difference between air flow rates over the two wing surfaces, and to obtain sufficient lift small insects must flap their wings very fast indeed. Many insects turn their wings to get lift both on the upstroke and the downstroke, and increase air movement by bending the wing along its length and suddenly releasing it, or by clapping the wings together dorsally and then flinging the leading edges apart (Figure 15.2e). Hovering, when there is no air movement provided by the whole body moving forward, is the most exacting form of flight. Lift is acquired only after several strokes and even then it is only obtained near the bottom of the downstroke and lost again as the wing turns for the upstroke.

Recent analysis emphasises that the forces acting on a wing that stops and starts repeatedly as it beats up and down are different from those on a wing moving at constant speed. Air over the leading edge forms a vortex bound to the upper surface of the wing and spiralling out towards the wing tip (see Figure 15.2*f*). This is the main aerodynamic mechanism enabling small insects to fly.

15.3.3 Methods of flapping

Insect wings are raised by contraction of indirect dorsoventral muscles that pull down the roof of the thorax (Figure 15.3*a*). For the wing downstroke, muscles are of three kinds:

– **Direct muscles** attached to the wing bases, pulling them down. These occur in dragonflies but in few other insects.
– **Indirect muscles**: in most insects, changes in shape of the thorax move the wings on the downstroke also, by contraction of the anteroposterior muscles (Figure 15.3*b*).
– **Asynchronous ('myogenic') muscles**: these are indirect muscles found only in Diptera, Hymenoptera and some small beetles and bugs. This form of striated muscle is able to contract much faster than nerves can conduct, being stimulated by mechanical stretch, usually caused by an antagonist muscle. The muscles depend on nerve impulses for background stimulation, and rely on energy stored in resilin in the wing hinge. These are the fastest, most elastic and most isometric muscles known, enabling up to over 1000 wing beats per second.

15.3.4 Control of flight

The wing-flapping rhythm is set by groups of interneurons constituting the 'central pattern generator', situated in the thoracic ganglia. The brain receives sensory input from eyes, antennae and mechanoreceptors placed at the base of wings, the whole head (dragonflies) or in Diptera the halteres, modified hind wings that beat up and down with the fore wings, any asymmetry of strain being detected by the receptors at the haltere base. The brain receives and coordinates the sensory input and may change the flight rhythm. Correction cannot be instantaneous in any insect, since the wing moves too

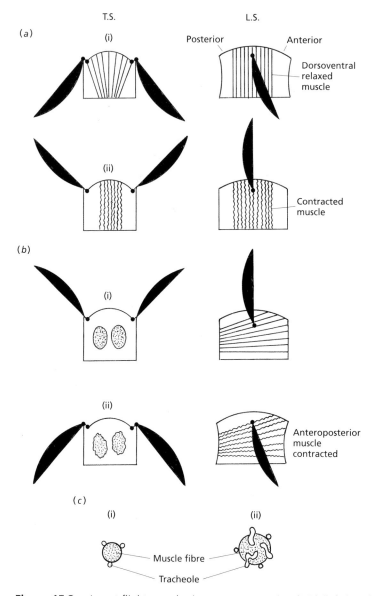

Figure 15.3. Insect flight muscles in transverse section (T.S.) (left-hand column) and longitudinal section (L.S.) (right-hand column): (a) in upstroke, for all insects (i) before and (ii) after muscle contraction; (b) in downstroke, for most insects (i) before and (ii) after muscle contraction; (c) T.S. muscle fibres with tracheoles (i) in a dragonfly and (ii) in a fly.

fast, and the only changes that the insect can make are to alter the angle of attack or the stroke plane of the wing, or to stop flying.

15.3.5 The metabolic cost

The energetic cost of flight is enormous. Oxygen is never limiting for an insect: the tracheal system can supply enough oxygen sufficiently fast to fuel the smallest flapping insect. Oxygen diffuses faster in air than in water by a factor of 10000, and with a primary tracheal trunk up to 0.74mm long (as in locusts) no energy need be spent on ventilation movements. Tracheoles may even indent muscle fibres (Figure 15.3c), which can be as large as 500 μm in diameter. By contrast, activity in vertebrates is limited by the rate at which oxygen can be supplied from blood capillaries (the diameter of the muscle fibres does not exceed 20 μm, exactly as in dragonflies where there is no tracheolar indentation). Vertebrates often need to draw on anaerobic respiration, where food is oxidised only as far as lactic acid formation and an oxygen debt is built up. Insects never need to be anaerobic (and lactic acid would flood the small body). Food supply is the limiting factor for an active insect. The blood that supplies food is not circulated fast, and muscle contraction squeezes it away from the sites where it is needed. However, insect blood can contain large and variable quantities of food. It is 1–3% carbohydrate, with similar quantities of fat and about 2% amino acids. The sugar is mostly trehalose, a disaccharide that can be tightly packed, rather than the monosaccharide glucose as in vertebrates, Mammalian blood by contrast contains small and precisely regulated quantities of foods (0.1% sugar and 0.1% fat, with very little amino acid except in the vessel between the intestine and the liver). This is necessary to maintain the stability of the respiratory pigment haemoglobin, not present in insects.

Only adult insects can fly. Some adults are wingless or seldom fly, and all other stages in the life cycle rely on other means of locomotion such as walking, crawling, swimming or burrowing.

15.4 What is distinctive about insect life cycles?

15.4.1 The egg

The egg is often very resistant and may be an important resting stage. Its early development is described in Chapter 19.

15.4.2 Division of labour

Within the life cycle, division of labour is made possible by metamorphosis, as in crustaceans: many insects have a feeding larva different in form and food supply from the reproductive adult. Unlike crustaceans, however, larvae are not the agents of dispersal if the adults can fly.

15.4.3 Exopterygotes

These are insects that have a gradual metamorphosis from the larva (sometimes called a 'nymph') into the adult ('imago'). Wing buds grow externally on the larval thorax (hence 'Exopterygota'). Many exopterygote adults and larvae such as those of cockroaches look very similar, live in the same environment and presumably compete for resources. Aquatic larvae (as in stoneflies and dragonflies) differ greatly from the adults in feeding mechanism and appearance. In mayflies the aquatic larva lives for up to three years and the very short-lived aerial adult has no gut and cannot feed at all.

15.4.4 Endopterygotes

These have 'complete metamorphosis' with total reorganisation of the body within the pupa. From the outset the larva contains the rudiments of adult organs (the 'imaginal discs'). These remain inactive during larval life but at the final moult when, in the absence of juvenile hormone, the pupa is formed, the adult rudiments develop and the larval tissues are resorbed. Endopterygote larvae typically look very different from the adults, have different feeding methods and exploit different niches

15.4.5 Diapause

Stages of inaction in the life cycle are often necessary to match an animal to its seasonal environment. The pupa of endopterygote insects may appear to be a resting stage while there is great activity inside, but then it may become entirely quiescent and provide a resting stage in the life cycle, additional or alternative to the egg. 'Diapause' is a period of arrested development, characterised by low oxygen consumption. The cue to enter diapause is most frequently a change in day length ('photoperiod'), rather than in temperature or humidity, which may be very variable. In *Pieris rapae*, for example, the shorter day length at the end of summer switches off brain hormone in the final larval stage. The pupa can be formed, but cannot moult and therefore goes into diapause. After an extended period of chilling (winter) the brain hormone switches on again, the pupa moults and the butterfly emerges.

15.5 What are the main Orders of insects?

The enormous numbers of insect species are assigned to many Orders, usually differing in life histories and in characteristic feeding methods. Like the different phyla of the Animal Kingdom, Orders are distinguished by recognisable basic forms (of adult or larva) showing adaptation to particular niches. There are two major divisions of insects:

I. **Apterygota**: some insects have lost their wings, but Apterygota are primitively wingless hexapod arthropods: an assemblage of forms not closely related, possibly having evolved separately from each other and from Insecta. See examples in Figure 15.4*a*.

II. **Pterygota**: insects primitively winged. There are two groups:

1. **Exopterygota** (Figures 15.4 and 15.5) with gradual 'incomplete' metamorphosis and the wings developing externally as larval wing buds.

   A. **Palaeoptera** are those exopterygotes in which the wings are not folded over the abdomen when at rest but are held out sideways or up above the insect's back. There are two-present day Orders, both with aquatic larvae:

   – **Odonata** (see Figure 15.4*b*): dragonflies are active predators with biting mouthparts, small antennae, very large eyes and a long abdomen extending straight out posteriorly. The wings at rest are held vertically in the delicate 'damselflies' and horizontally in the larger true 'dragonflies' (the name may be reserved for this subdivision or given to the whole Order). They are ancient insects whose uniform structure has changed remarkably little since the Carboniferous. Fossil insects from this era with a wing span of about 0.6 metres are the largest insects known.

   Their mating system is unique (Figure 15.4*c*). The male usually patrols a territory beside water, captures a female and flies 'in tandem' with her. He fills his accessory genitalia under the front

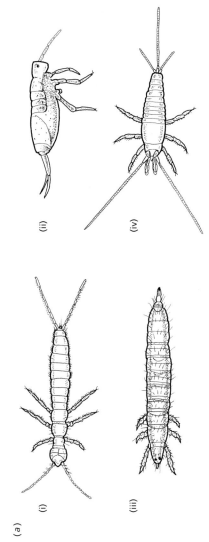

(a)

(i)

(ii)

(iii)

(iv)

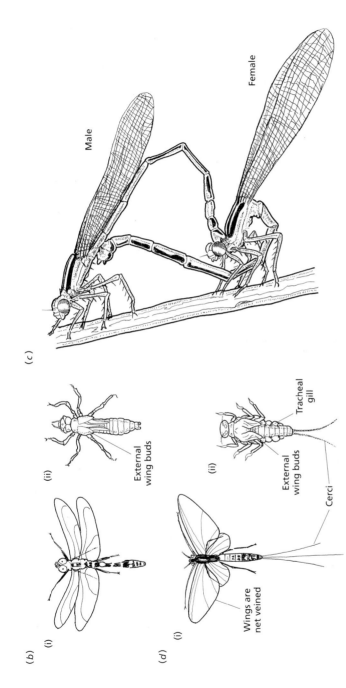

Figure 15.4. The diversity of insects. Apterygotes and palaeopteran exopterygotes (Odonata and Ephemeroptera): (a) apterygotes, (i) dipleura, (ii) springtail (length about 1 mm), (iii) protura (length 7 mm), (iv) bristletail (length 2 cm); (b) anisopteran dragonfly, (i) adult and (ii) aquatic larva; (c) two damselflies in tandem; (d) mayfly, (i) adult and (ii) aquatic larva.

end of the abdomen with sperm, cleans out from the female any sperm from a previous mating, and then inserts his own sperm. Eggs are laid in water or on submerged plants. The aquatic larva seizes prey with a 'mask'; it may live in water for several years.

— **Ephemeroptera:** mayflies: Figure 15.4*d* shows the adult and larva. There is also a pre-imago stage which (unlike all other immature insects) has wings.

B. **Neoptera** are those insects in which the wings are normally folded over the dorsal surface of the abdomen when at rest. They include:

— **Dictyoptera** (Figure 15.5*a*): cockroaches and mantids.
— **Orthoptera** (Figure 15.5*b*): locusts, grasshoppers and crickets.
— **Phasmida:** stick insects and leaf insects have highly adapted forms, as their names indicate. By day these nocturnal predators are effectively concealed among twigs and leaves.
— **Plecoptera** (Figure 15.5*c*): stoneflies.
— **Dermaptera** (Figure 15.5*d*): earwigs They are semi-social and make good mothers – and do not inhabit the ears of any animals.
— **Isoptera** (Figure 15.5*e*): termites are social insects sometimes called 'white ants', being colonial and polymorphic. They feed on wood and may be very destructive to buildings; they construct and inhabit the large often pointed mounds seen in the tropics.
— **Hemiptera** (Figure 15.5*f*): bugs; a term misapplied to other forms.

There are other smaller Orders (Figure 15.6*e–k*).

— **Anoplura:** sucking lice.
— **Mallophaga:** biting lice, ectoparasitic on birds and mammals.
— **Psocoptera:** book lice, feeding on fungi, for example those growing on the bindings of damp books.
— **Embioptera,** web-building insects living under bark in tropical countries.
— **Thysanoptera:** thrips or thunderflies, tiny insects fringed with hairs. The mouthparts are asymmetrical.
— **Grylloblattoidea,** which have been thought to be remnants of the primitive stock from which Dictyoptera and Orthoptera evolved.

2. **Endopterygota** (see Figure 15.6) with complete metamorphosis and a pupal stage, within which adult structures including wings develop from imaginal discs.

There are four major large Orders and a number of smaller ones:

— **Lepidoptera** (Figure 15.6*a*): butterflies and moths. A very large and well known Order of insects, united by having scaly wings. Lepidoptera are among the most recent insects: they arose in the Tertiary and evolved in close relationship with flowering plants. The larvae (caterpillars) feed on leaves, stems or roots and the adults feed by sucking nectar from flowers. Large diurnal butterflies owe their conspicuous and often beautiful coloration to the scattering of light by the scales on their wings. Most families of Lepidoptera are small nocturnal moths. Some primitive 'Microlepidoptera' feed using biting mouthparts, but in all others the mandibles disappear and the maxillae are enormously extended to form a sucking 'proboscis', coiled under the thorax when not in use.
— **Hymenoptera** (Figure 15.6*b*) have small hooks joining the posterior edge of the fore to the hind wing on each side ('married' wings). This diverse Order includes

    (a) **Symphyta**: sawflies, with larvae looking like caterpillars and adults with no obvious 'waist' and
    (b) **Apocrita** with grub-like larvae and adults with a constriction between the thorax and abdomen. These include:

        (i) **Parasitica**: ichneumonid flies, gall wasps, which lay eggs in other insects or in plants; the larvae live as parasites inside these hosts (and may be important regulators of the population sizes of some of our pest species).
        (ii) **Aculeata**: bees, wasps and ants, with adults larger than in Parasitica and with the female ovipositor modified as a sting. All provide food for their helpless larvae, in the form of pollen and honey (bees), insect prey (wasps) or fungi cultivated on plant material (some ants). There are many species of solitary bees and wasps but most of them, and all ants, are social insects (see below).

— **Coleoptera** (see Figure 15.6*c*): beetles are characterised by 'elytra', fore wings modified into hard resistant cases under which the membraneous hind wings are folded. When not in flight a beetle appears to be completely covered in armour. All beetles have strong biting mouthparts. Beetles are first found in the Permian as a very compact and isolated group of insects: today they constitute a uniquely vast and diverse Order, with about half a million species

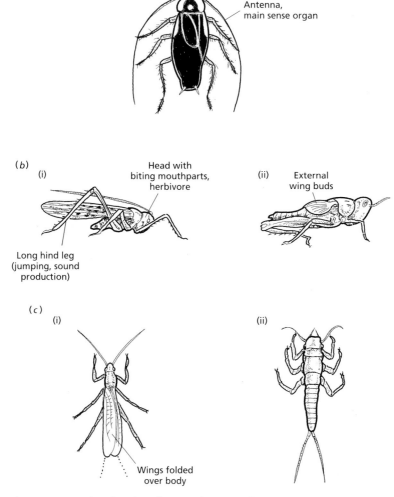

(a) Antenna, main sense organ

(b) (i) Head with biting mouthparts, herbivore

(ii) External wing buds

Long hind leg (jumping, sound production)

(c) (i)

(ii)

Wings folded over body

Figure 15.5. The diversity of insects (continued). Neopteran exopterygotes: (a) Dictyoptera, cockroach adult (larva, very similar, not shown); (b) Orthoptera, locust, (i) adult and (ii) larva, showing externally developing wing buds; (c) Plecoptera, stonefly, (i) adult, found on stones near water and (ii) aquatic larva;

ranging from 0.5 mm to 21 cm in length. The larvae are typically legless grubs but some are predators with jaws and legs.

– **Diptera** (see Figure 15.6d) : true flies have only one pair of wings. The hind wings have become modified into halteres, which are sensory structures assisting flight (see above). Flies are relatively

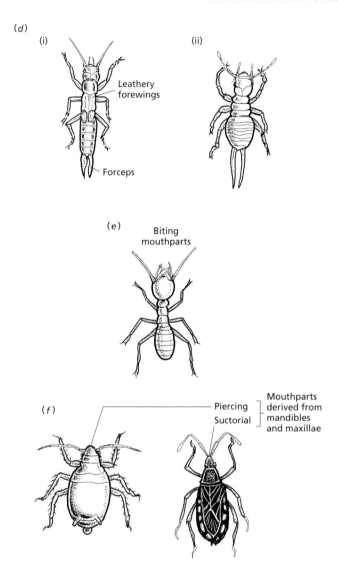

Figure 15.5 (cont.). (d) Dermaptera, earwig, (i) adult and (ii) larva; (e) Isoptera, termite worker (soldier); (f) Hemiptera, bugs, adults of the two suborders, feeding by piercing or sucking plants or animals.

recent insects that arose in the Cretaceous, in association with flowering plants. There are very many diverse species including insects with piercing and sucking mouthparts (e.g. midges, mosquitoes, craneflies, horseflies) and the great majority (e.g. house

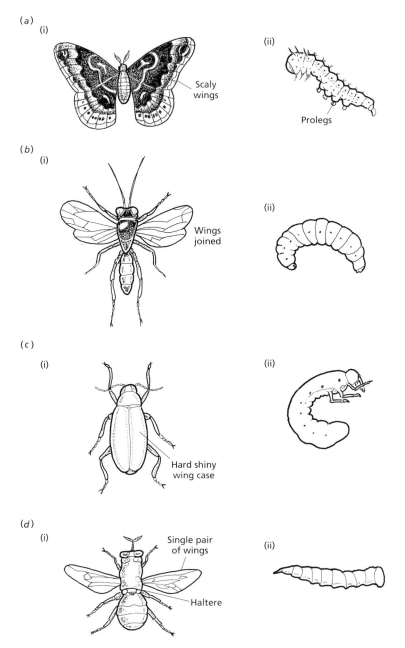

Figure 15.6. The diversity of insects (continued). Endopterygotes, generalised examples: (a) Lepidoptera, a butterfly, (i) adult and (ii) larva; (b) Hymenoptera, a bee, (i) adult and (ii) larva; (c) Coleoptera, a beetle, (i) adult and (ii) larva; (d) Diptera, a fly, (i) adult and (ii) larva; and (i) adults and (ii) larvae of three smaller Orders;

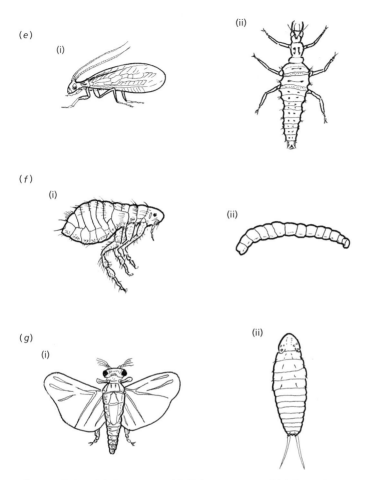

Figure 15.6. (e) Neuroptera, (f) Siphonaptera and (g) Strepsiptera.

flies, fruitflies), which have mouthparts specialised for sucking only. Many dipteran larvae are headless maggots living in soft substrata.

Smaller Orders of endopterygotes include:

- **Trichoptera:** caddis flies, a small group resembling moths but with hairs on their wings rather than scales. There is a prolonged aquatic larval stage, when the larva usually builds a characteristic case of assorted fragments of stones etc. from which its head projects for feeding.
- **'Neuroptera'** (Figure 15.6e): alder flies, lacewings, and ant-lions are not a natural group but a combination of endopterygote Orders

where both adults and larvae are predators with biting mouthparts. The wings look like nets of gauze and are held tent like over the abdomen; flight is rather feeble.

— **Mecoptera**: scorpion flies, first found as fossils in the Lower Permian, are thought to be survivors of the primitive stock from which evolved the 'Neuroptera', Lepidoptera, Trichoptera and Diptera.

— **Siphonaptera** (Figure 15.6*f*): fleas are laterally compressed and wingless ectoparasites that pierce the host with stylets and suck blood.

— **Strepsiptera** (Figure 15.6*g*), much modified parasites of some bugs and Hymenoptera.

15.6 How could social behaviour have evolved?

Ants, bees and wasps among the Hymenoptera (and termites, separately evolved in the exopterygote Isoptera) not only live in colonies but show remarkable division of labour among the members of those colonies. The biology of social insects is fascinating, but only one question is considered here: how could this behaviour have evolved? This is not a problem about lack of intermediate stages: there are, for example, many solitary bees showing rudiments of social behaviour. The problem is to identify selective advantage for apparently altruistic behaviour. Typically there are very few reproductive individuals: hymenopterans have only a single queen per colony and the males (drones) die after the nuptial flight, during which she collects sperm that lasts for the rest of her life. The workers (and soldiers in ants) are all sterile females. In termites, each colony has a queen, who grows enormous, and a fertile king, who lives with her and fertilises her eggs as they are produced, and the workers are sterile males or females. How could it benefit the individual workers to cooperate in the rearing of their siblings and work to maintain the colony, instead of using their own reproductive potential? Natural selection acts on individuals, not on species as a whole.

The problem of altruism disappears when the focus is shifted to the genes. Natural selection acts upon individuals, but the effect of its action is that some groups of genes rather than others will be passed down the generations (see Chapter 1). If two individuals are closely related, the reproduction of either may be equally advantageous for transmission of a given gene. There is in this context nothing special about offspring. When siblings are as closely related to each other as to potential offspring, their genes are as likely to be transmitted when all siblings promote the reproduction of one of them. In Hymenoptera a special mechanism makes sibling workers even more closely related to each other than to potential offspring, thereby increasing the advantage of their social behaviour.

15.7 Why has study of the fruit fly *Drosophila* been so important?

Drosophila was T. H. Morgan's inspired choice for animal breeding experiments in the earliest years of the twentieth century. These fruitflies (*Drosophila melanogaster*) were already in 1901 being cultured for embryological work in the laboratory of the University of Columbia, USA. They are ubiquitous in the wild and easily obtained and cultured; they are small insects that produce large numbers of offspring in a very short time, and may complete their life cycle in two weeks.

15.7.1 Early work by Morgan and his group

Inbreeding and outbreeding experiments confirmed that Mendel's Laws (see Chapter 1) applied to fruit flies as much as to garden peas. Work on *Drosophila* revealed the phenomena of linkage and crossing over, showed that genes were arranged longitudinally along chromosomes and related the position of genes to the frequency of the breaking of linkage. The job was relatively easy in *Drosophila*, since there are only four pairs of chromosomes (later work used a different species, *D. pseudobscura*, where the chromosomes can be seen more clearly). *Drosophila* was a fortunate choice, since the salivary glands in the larva have giant 'polytene' chromosomes made from many strands of the genetic material in a bundle. These chromosomes were large enough to be accessible to experiment even in Morgan's time: under the light microscope, transverse bands could be seen at different intervals, so that different parts could be recognised, as could 'puffs' when the genes were active. Genetic mapping was checked and extended by experiments in which parts of the chromosomes were cut out, turned round or rearranged.

T. H. Morgan, A. H. Sturtevant, H. J. Müller and C. B. Bridges in 1915 published *The Mechanism of Mendelian Heredity*. The merging of cytology and genetics into a single science was the first major contribution of the *Drosophila* work of the Morgan school.

15.7.2 The evolutionary role of mutations

Further work by Morgan's group and by others such as T. Dobzhansky, and (with a gift of cultures from Columbia) S. S. Chetverikov in Russia, in the 1920s established the connection between Mendelism and Darwinism. At that time genetics and evolution were viewed very separately, with much controversy about their relationship. H. De Vries working on Evening Primroses (*Oenothera biennis*) observed sudden large heritable changes that he termed 'mutations': these were generally deleterious and could not be seen as the raw material of evolution. Morgan redefined mutations as small changes in genes, initially recessive but retained in the genotype, and showed how such mutations could be the source of variation acted on by natural selection. For the study of mutations, *Drosophila* was a rewarding subject, having the complex insect structure with many characters under direct genetic control. A large 'library' of mutations was soon acquired, ranging from large scale malformations to patterns of bristles on the cuticle. Mutations were shown to occur in wild populations also, to be imitable by heat shock at critical points in development (R. Goldschmidt) and to be caused by ionising radiation (Müller). A balanced polymorphism in the population was described by M. Teissier; the statistical approach to populations by R. A. Fisher was especially important. Genes were revealed as large and intricate entities that interacted with each other and with the environment, and genetics became part of the 'evolutionary synthesis' of the 1930s and 1940s.

15.7.3 Speciation

Speciation in *Drosophila* is marked, most spectacularly in Hawaii where 450 species have been described. New Hawaiian islands are continually being formed and are biologically 'empty' at first. Small insects are weak fliers and colonists are few: these speciate readily as each island presents a great diversity of habitats, sexual selection is strong and lava flows may further divide populations. Many closely related species of *Drosophila* are therefore available for comparison; often there is remarkably little genetic difference between them.

15.7.4 Modern genetics

Accumulated knowledge of *Drosophila* species and mutants has provided an invaluable basis for application of modern knowledge and techniques concerning the nature and action of genes. Along with the less elaborate forms *Escherichia coli* (a bacterium) and *Caenorhabditis elegans* (a nematode), *Drosophila melanogaster* has now been chosen for mapping of the entire genome.

16

Animals with lophophores

Sessile animals require special structures for food collection: the stinging cells on the tentacles of cnidarians and the feeding methods of sponges and barnacles have already been discussed. The present chapter introduces four groups of animals with lophophores. As adults all are sessile, with poorly developed heads, U-shaped guts and a protective outer covering over what is most often a colony of zooids. A superficially similar phylum is also introduced for contrast, and the relationships of animals with lophophores are discussed.

16.1 What is a lophophore?

A lophophore is a circular or horseshoe-shaped fold of the body wall that encircles the mouth and bears numerous ciliated tentacles, the anus being outside the tentacular ring. The tentacles are hollow, each one containing a branch of the coelom derived from the 'mesocoel', which is the name given to the middle of three embryonic coelomic cavities. Lateral cilia on the tentacles draw water bearing food particles into the tentacular ring, frontal cilia convey it down each tentacle and it goes out at the base, where food is trapped (Figure 16.1a).

16.2 Which animals have lophophores?

Three phyla, the Phoronida, Bryozoa and Brachiopoda, with lophophores exactly fitting the above description are frequently united as 'Lophophorata'.

16.2.1 Phoronida (see Figure 16.1*b*)

Fourteen species are known, in two genera *Phorona* and *Phoronopsis*. They are up to 200 mm long. All are marine. The adult is solitary, worm-like and able to move only within the chitinous tube that fastens it to the shallow sea bottom. The lophophore is either simple with rather few tentacles, or elaborate and spirally coiled. There is a closed blood system with haemoglobin in corpuscles, the nervous system is subepidermal and largely diffuse. The animals are mostly hermaphrodites, releasing gametes through the excretory pores. The fertilised eggs develop into free swimming 'actinotroch' larvae.

16.2.2 Bryozoa (see Figure 16.1*c*)

The Bryozoa have 20000 species of which 5000 are living today. They are colonial, with zooids about 0.5 mm long, and most are marine. Bryozoa ('moss animals') have also been called 'Polyzoa' (because there are many individuals in a colony) or 'Ectoprocta' (because, as with all true lophophores, the anus is outside the tentacular ring). Bryozoa are sessile colonies of miniaturised animals. Microscopic examination shows that each inhabits a separate little 'box' of cuticle within the colony, which at once distinguishes them from hydroid cnidarians, and their high speed of withdrawal into their boxes is a further distinction. The cuticle is made of chitin and protein, sometimes also calcium carbonate. The zooid body typically has a stationary trunk and an eversible lophophore: a special opening appears when muscular contraction increases the fluid pressure in the coelom. There is a simple nerve ganglion giving off nerves, a U-shaped gut, and (in such small animals) no excretory nor blood systems. The colonies grow by budding, or the zooids may contract down with regression of the gut and lophophore, leaving remnants as 'brown bodies' when most of the tissue regenerates. The zooids are hermaphrodites, the gametes being shed into the coelom and passing out through pores at the tentacular base. The fertilised eggs are usually brooded, and hatch as free-swimming larvae.

Division of labour between the members of a bryozoan colony commonly occurs. In addition to feeding zooids (described above), there may be 'bird-like' individuals with jaws serving for defence (Figure 16.1*d*) and 'whip-like' ones that clear away waste and deter settlers.

Bryozoa are very common intertidal animals, growing either as mats covering seaweed or as separate tree-like tufts. In addition to the main marine class, there is one class that is entirely freshwater and a third that is known mainly as fossils, very abundant since the late Cambrian.

(a)

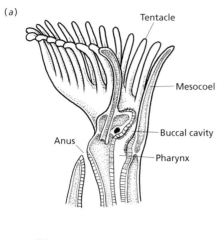

Tentacle

Mesocoel

Buccal cavity

Anus

Pharynx

(b)

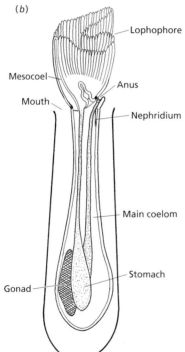

Lophophore

Mesocoel

Anus

Mouth

Nephridium

Main coelom

Stomach

Gonad

Figure 16.1.   (a) A lophophore; (b) longitudinal section *Phoronis*.

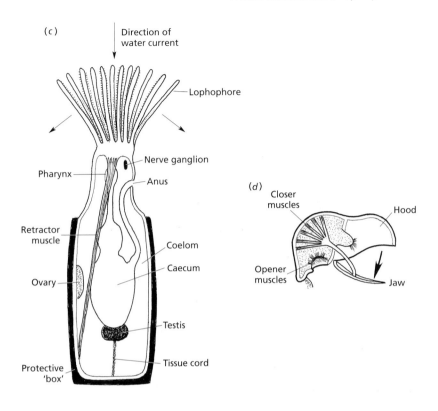

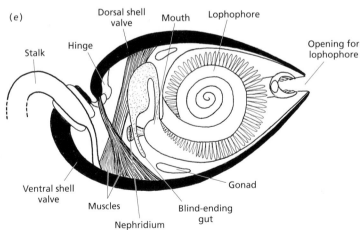

Figure 16.1 (*cont.*). (*c*) A generalised bryozoan zooid; (*d*) a 'bird-like' zooid of a bryozoan colony; (*e*) a brachiopod with the lophophore coiled internally.

16.2.3 Brachiopoda (Figure 16.1e)

There are about 13000 species, of which only 350 are living today. The shell is less than 100mm long or wide. All are marine. Brachiopods are solitary sessile animals enclosed by the dorsal and ventral valves of a shell. The two halves of the shell are unequal and the ventral half bears a stalk that anchors the animal to the substratum. Resemblance to the bivalve molluscs, where the two halves of the shell are lateral and equal, is only superficial. The main body of the brachiopod consists of a coelom surrounding the gut, excretory metanephridia and either testes or ovaries as the sexes are separate. There is an open blood system and a ganglionic nerve ring round the oesophagus. Contraction of well-developed muscles moves the shell valves. The body lies posteriorly in the shell and the body wall extends forward, folded as a pair of mantles lining the shell. The horseshoe shaped lophophore develops as a pair of arms that may become highly elaborate. It is coiled within the mantle cavity, from which it can be extruded.

Brachiopods are extremely common in the fossil record, from the early Cambrian onward. The stem groups reach their peak of diversity in the Ordovician and the greatest number of genera is found in the Devonian. They are later largely displaced by molluscs. Today they are still very widely distributed in the seas, although they are seldom common at any one place or depth.

There are two very different brachiopod groups, which may not be closely related. The Articulata (e.g. *Terebratula*) have calcareous shells with valves interlocking by teeth on one valve fitting into sockets on the other, and the dorsal valve usually has calcified outgrowths supporting the lophophore. The anus is lost in most living forms. The Inarticulata (e.g. *Lingula*) have organophosphatic shells with no articulation and no outgrowth. The valves are held together by complex musculature and there are other internal differences from Articulata. The earliest fossil brachiopods are inarticulate. *Lingula* is an outstanding example of a species which has persisted unchanged from the Ordovician to the present day.

16.2.4 Pterobranchiate Hemichordata

In addition to the three above-mentioned phyla, pterobranchs also have lophophores exactly fitting the above description, except that the tentacles do not completely encircle the mouth (see Figure 16.2a), but the mesocoel is continuous underneath the tips of the tentacular horseshoe. Yet pterobranchs are members of the phylum Hemichordata (see Chapter 18).

16.3 Are animals with lophophores protostomes or deuterostomes?

This question is not easily answered, since phoronids, bryozoans and brachiopods present a mixture of the characters introduced in Box 5.3:

16.3.1 Phoronida

Like protostomes, the blastopore becomes the mouth, yet like deuterostomes cleavage is radial, early development is regulative and the coelom has three primary divisions. Further protostome-like characters are the protonephridial excretory tubules in the larva, the nervous system inside the epidermis and the tube made of chitin. The mesoderm arises from the gut wall as in deuterostomes but the coelom is formed by splitting as in protostomes.

These contradictory characters have in the past been reconciled by placing phoronids as a basic group from which protostomes and deuterostomes have diverged, but further knowledge (such as molecular evidence) now makes such a theory very difficult to sustain.

16.3.2 Bryozoa

At first sight bryozoans are good deuterostomes, with radial cleavage, the blastopore forming the anus and coelom formation from three original cavities, although development is unusual in that the lophophore and gut bud from within the larva. Yet they have a number of resemblances to protostomes, notably the presence of chitin. Such resemblances might be dismissed but for molecular evidence: 18SrDNA analysis places bryozoans among the protostomes, not too distant from phoronids, annelids and molluscs.

16.3.3 Brachiopoda

Their development does not fit neat categories. Like deuterostomes, they have radial cleavage and regulative development. The blastopore forms the anus in some but not all species, and the coelom originates in three parts. Mesoderm formation, however, is enterocoelic only in the Articulata. In the Inarticulata, the more primitive group, mesoderm proliferates from the gut wall and then splits, just as in phoronids; apparently methods of coelom formation are more plastic than has been assumed. Brachiopods like bryozoans are increasingly seen as protostomes, on account of both fossil and molecular evidence (see Chapter 20). Whether brachiopods are monophyletic remains uncertain.

16.3.4 Pterobranchiates

These organisms remain classical undoubted deuterostomes by every criterion available, developmental, morphological and molecular.

16.4 What are the relationships of Entoprocta?

Entoprocta (Figure 16.2*b*) have 150 species, of length 0.5–5 mm. Most are marine with a few freshwater species. They may be solitary (e.g. *Loxosoma*) or colonial (e.g. *Pedicellina*). They are superficially similar to 'Ectoprocta' (Bryozoa), being small, sessile, filter-feeding animals, mostly colonial and marine. However, they lack a coelom (they may be called acoelomate or pseudocoelomate, having many gelatinous cells in what might otherwise be a cavity). The tentacles are solid, borne on a collar that is not retractable but can be folded down, and the anus is within the tentacular ring. A further contrast with lophophores is that the water currents flow in the opposite direction, as lateral tentacular cilia draw water in at the base of the tentacular ring, up each tentacle and out at the tip, while frontal cilia trap food particles and convey them down to the mouth.

Entoprocts have no exoskeleton. There is a single protonephridium, no respiratory nor blood systems, and nerves arise from a single ganglion. The animals are hermaphrodites and the egg hatches as a trochophore-like larva.

Entoprocts are undoubtedly protostomes, with spiral cleavage, the mouth formed from the blastopore and mesoderm developed from the 4d cells (it does not split, there is no body cavity). Analysis 18SrRNA analysis supports this placement among the protostomes, not particularly close to the bryozoans nor to other animals with lophophores.

The only phylum apparently close to the entoprocts was discovered in 1995 by P. Funch and R. M. Kristiansen: **Cycliophora** (Figure 16.2*c*). This phylum is known only from one species, *Symbion pandora*, 0.35 mm in length, living on the mouthparts of the marine Norway lobster *Nephrops norvegica*. The life cycle is complex. The sessile feeding stage has a funnel-shaped crown of tentacles, and water passes from the base to the tips of the tentacles as in entoprocts. The animal is solitary, enclosed in a sculptured cuticle resembling that of some

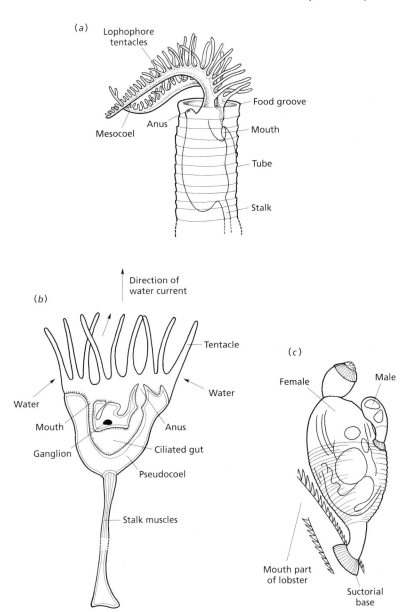

Figure 16.2. (a) *Rhabdopleura,* a pterobranch hemichordate; (b) a generalised entoproct; (c) *Symbion pandora,* a cycliophoran.

nematodes, with a stalk for temporary attachment. This stage is female, with a small male attached. After fertilisation, development is that of a typical protostome and the larva is like a trochophore, except that internally it buds off a clone of individuals, which settle to form the feeding stage.

16.5 Should there be a group called Lophophorata?

Should the animals with lophophores be combined in one group called 'Lophophorata'? Phoronida, Bryozoa and Brachiopoda may be fairly closely related, and on present evidence appear to be protostomes. However, the pterobranchiate Hemichordata are classical deuterostomes by every criterion available, developmental, morphological and molecular. This is most readily explained as a remarkable example of convergent evolution of the feeding structure, and to combine such distantly related animals as 'lophophorates' would therefore be misleading.

Box 16.1
How animals feed

Animals cannot manufacture organic foods from their inorganic constituents, and therefore they are dependent on green plants or microorganisms for their food. As land animals we are used to the assumption that our food chains begin with photosynthesis by green plants, but in the sea this is relatively rare. Originally the sea contained single-celled organisms only, and these are still the basis of many marine food chains. Light does not penetrate far and photosynthesis can occur only very near the surface; detritus falling from the surface layers becomes the basis of many food chains in deeper waters and on the sea bottom. Chemosynthetic bacteria are important, in particular the sulphur-based bacteria in deep sea vents (this region is only now being explored). Small animals, or those with a large food-collecting area, can exist on a purely microbial diet.

Small particle feeding

Suspension or 'filter'-feeding

Filter-feeding can occur only in water. It is very common in marine larvae, and most phyla have examples of adult animals extracting food particles either from water through which they move or from a current of water passed over a food-collecting surface. Such a surface frequently is covered in cilia, usually two sets at right angles, one set producing the current and the other trapping food particles, which may be bound in secreted mucus before being conveyed to the mouth.

Ciliary feeding occurs, for example, in tubicolous polychaetes, bivalve molluscs and animals with lophophores. Note that 'filter-feeding' is a misleading description of ciliary feeding, because cilia are too small and too closely packed for water to pass between them. Arthropods have no cilia, but cuticular hairs on feeding appendages may trap small particles from water.

Detritus feeding

Ingestion of particles of decomposing organic matter and microbes contained within it (detritus) was probably the ancestral feeding method and remains an important basis for aquatic food chains.

Large particles or food masses

These are needed by most large animals.

Deposit feeding

The consumption from the substratum of organic detritus is called deposit feeding. Detritus may be swallowed whole into the animal's body where food is extracted, and (for example, in *Arenicola* and earthworms) the familiar worm casts are left behind.

Intake of plant food

Plant food may be obtained by browsing and grazing or by scraping and boring. Tools are usually required, such as the molluscan radula or the jaws of arthropods and many worms, and chemical secretions may assist food gathering. Hard structures are necessary also to penetrate the cellulose cell walls. Digestion of plant food is difficult for animals, since they lack the enzymes that break down cellulose. Animals frequently retain symbiotic bacteria, fungi or protista in a special gut compartment. The ruminant stomach is a familiar example: many invertebrates that feed on seaweed (e.g. sea urchins, a few littoral molluscs) or land plants (e.g. many arthropods and molluscs) similarly rely on microbial symbionts for breakdown of cellulose.

Living animals

Live animals may be the prey either of resident parasites (see Box 6.1) or of carnivores, which kill and eat their prey. Examples of predators occur in all phyla (there are even improbable carnivores among sponges and ascidians). Attack

Box 16.1 *(continued)*

frequently requires poison (e.g. in cnidarian nematocysts, nemertine proboscides, arachnid poison glands) as well as, or instead of, hard structures.

Fluid feeding

Piercing and sucking

Plant or animal bodies may be pierced or sucked by parasites or by other predators.

Uptake of organic food in solution

Soluble organic food may be taken up over any permeable body surface and this is now known not to be confined to endoparasites but to be widespread in aquatic, particularly marine, organisms. It is usually a secondary development and an auxiliary feeding method, but some animals without guts (e.g. pogonophorans) can feed in no other way.

17

Echinodermata

Echinoderms include starfish, brittle stars, sea urchins, sea cucumbers, sea lilies and feather stars. They are very unusual animals. They are radially symmetrical with no head end, yet are far more complex than the cnidarians in that they are coelomates with three cell layers, and most can form a temporary brain that coordinates locomotion. Their size varies: starfish range from 1 cm to 1 metre in diameter, and some sea cucumbers are as much as 2 metres in length. About 7000 species have been described, all marine. While the number of species is not very great, echinoderms are extremely numerous as individuals in unpolluted seas and deep water. This is a successful and very interesting phylum.

The echinoderm body plan is not only unusual, it has some features unique among animals. Discussion of these features, and an indication of the directions of evolution that have occurred, is followed by an account of the larval forms; unlike the invertebrate phyla so far introduced, echinoderms are deuterostomes. In conclusion, the evolutionary history of the phylum can be outlined, largely because (unlike soft-bodied invertebrates) the echinoderms have left a well-marked sequence of fossils.

17.1 What is unique about echinoderms?

17.1.1 The water vascular system

The echinoderm water vascular system has a central coelomic cavity that transports sea water within the animal and bears the tube feet, thin-walled extensions of the epidermis lined with mesoderm and connected with the radial canals. In starfish, for example, sea water enters through a sieve plate (Figure 17.1a) and passes by way of a central 'stone canal' to a ring canal and then along radial canals and into the tube feet. The functions of tube feet may include respiration (echinoderm skin is impermeable), food collecting, acting as sensory tentacles and locomotion. The 'walking' tube feet of starfish and sea urchins are connected to an 'ampulla', a seawater reservoir that can be squeezed to extend the tube feet (Figure 17.1b,c). Not all can be extended at once as there would not be enough water inside the animal, nor can all be retracted. The tips of the tube feet adhere to the substrate by suction in a sea urchin, but in a starfish by a chemical reaction. Secretion by the tube feet give rapidly variable adhesion as the starfish walks, since adhesive acidic carbohydrates and proteins alternate with substances of different ionic content that detach them.

17.1.2 The endoskeleton

Underneath the epidermis, perforated calcareous ossicles are formed by mesodermal cells. Each ossicle is a single crystal and is separated from its neighbours by living tissue (Figure 17.1d,e). This system is very different from that characteristic of hard skeletons in other animals, where crystals are embedded in an organic matrix. It allows increase in size without moulting, it can be rigid or hinged and it can be serviced by living cells. External spines justify the name of the phylum ('spiny skins').

17.1.3 The nervous tissue

The nervous system is a diffuse nerve network condensed in places into nerve fibres but without a brain. This suggests a degree of coordination similar to that of cnidarians, yet groups of nerve cells can, as will be explained, become temporary coordinating centres.

17.1.4 Mutable connective tissue

The connective tissue of echinoderms is a most striking unique feature. It is capable of rapid and reversible change in stiffness, often within seconds, under nervous control. The mechanism is a change in the viscosity of collagen fibres in the connective tissue, caused by ionic movements that alter weak interactions between extracellular macromolecules. The process is controlled by calcium binding to neurosecretory cells. The result is that muscle tone can be altered without the expenditure of energy involved in muscular contraction. Examples include the stiffening of sea urchin spines, which can anchor the animal, the stiffening and relaxation of the body walls of starfish and sea cucumbers, the autotomy of arms in brittle stars and the maintenance of tone in the extended arms of sea lilies. It is a property that the human animal can only envy.

17.2 What is unusual, but not unique, about echinoderms?

In addition to their unique features, echinoderms differ from most other invertebrates in various ways.

17.2.1 Pentaradiate symmetry

Radial symmetry is a legacy of their evolution from sessile animals that collected food from all sides. Most present-day echinoderms are motile animals, modified accordingly. Why echinoderms usually have five rather than any other number of arms is not clear.

17.2.2 Absence of blood and excretory systems

Food and metabolic waste may be transported by the water vascular system; amoebocytes in the coelom branches may carry metabolites to any permeable area of skin, where also gaseous exchange occurs. Osmotic regulation is almost completely absent. The body fluid is not always osmotically identical to sea water, for example the fluid in the water vascular system of a starfish is slightly hypertonic to the sea, but differences are small. There are no freshwater echinoderms.

17.2.3 Deuterostome development

Cleavage is radial and regulative rather than spiral with early determination of cell fates, and the blastopore becomes the anus. The mesoderm is derived

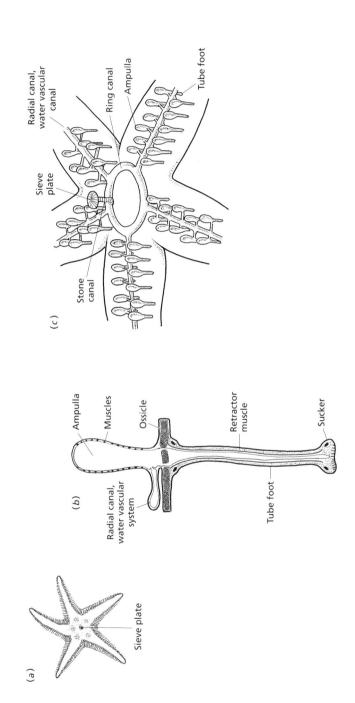

(a)

Sieve plate

(b)

Ampulla

Muscles

Ossicle

Radial canal, water vascular system

Retractor muscle

Tube foot

Sucker

(c)

Radial canal, water vascular canal

Ring canal

Ampulla

Tube foot

Sieve plate

Stone canal

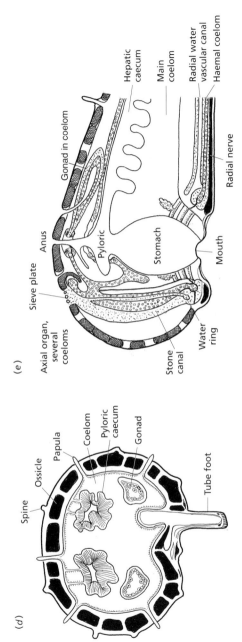

Figure 17.1. The anatomy of an asteroid starfish: (a) dorsal view of Asterias; (b) a tube foot with an ampulla; (c) the water vascular system; (d) transverse section of arm of Asterias; (e) vertical section of Asterias showing the many coeloms.

(d)

Spine
Ossicle
Papula
Coelom
Pyloric caecum
Gonad
Tube foot

(e)

Axial organ, several coeloms
Sieve plate
Anus
Gonad in coelom
Hepatic caecum
Main coelom
Radial water vascular canal
Haemal coelom
Pyloric
Stomach
Radial nerve
Mouth
Water ring
Stone canal

from the gut wall as the coelomic cavities are formed by a series of out-pouches (see Box 5.3).

17.2.4 The complexity of the coelomic cavities

Echinoderms are unsegmented, the coelom is not divided along a body axis as in an annelid, but it exists as a number of separate divisions persisting from its embryological origins. In addition to the main water vascular system, a starfish (for example) has a large perivisceral coelom, and haemal (misnamed), perihaemal and gonadal coeloms (Figure 17.1e).

17.3 How do different echinoderms feed and move?

All animals are constrained by their evolutionary history; in the echinoderms this constraint is particularly obvious, and can be related to many of their unusual features. The ancestors of the phylum were bottom-living sessile animals with massive calcareous skeletons. The mouth faced upward and food was collected from all sides (see above, section 17.2.1). Most modern echinoderm adults move slowly over the sea bottom, collecting their food into a mouth on the under surface. Figures 17.1 to 17.4 indicate the range of adult form in the six classes of modern echinoderms.

17.3.1 Echinoderm larvae

The larvae feed and move very differently from the adults. Most echinoderms have ciliated, filter-feeding larvae swimming in the marine plankton (see below).

17.3.2 Crinoidea (sea lilies and feather stars)

Although they may not be the oldest group of echinoderms, crinoids resemble their ancestors in being essentially sedentary suspension feeders collecting their food into an upwardly facing mouth (Figure 17.2a,d). Most of the sea lilies are extinct. Modern feather stars (e.g. *Antedon*, Figure 17.2b) hold out their much branched arms, which are stiffened by ossicles and mutable connective tissue, and collect food into the 'ambulacral groove' on the upper surface (Figure 17.2*c*). Here numerous tube feet bearing mucous papillae act like cilia in conveying the food (bound in mucus) to the central mouth. Feather stars are secondarily motile, being able to swim by waving the arms up and down.

17.3.3 Asteroidea (starfish, see Figure 17.1)

The Asteroidea constitute a large class with some 2000 species. They are predators or scavengers. Some starfish feed by sucking animals into the stomach but others evert the stomach as they feed: typically a starfish grips the two halves of a bivalve shell with the tube feet suckers and pulls the shell halves apart until it can insert the everted stomach through a tiny gap. The bivalve is then digested inside its shell and sucked into the starfish. Some species are suspension feeders using mucus and cilia.

Active predators must not only move, they must move in a directed way from place to place. How can the radially symmetrical starfish move in one direction, and without a brain how can it coordinate the stepping of the tube feet in different arms? Observation shows that any arm can become the leading arm, with the other four arms cooperating. Experimental cutting of the nerve ring in two places shows that tube feet in the arms beyond the cuts no longer cooperate with the leading arm, but step towards the tip of their own arm (Figure 17.3*a,b*). In an isolated arm, the tube feet step in coordination towards the tip of the arm only if the junction between the radial nerve and the nerve ring is present (Figure 17.3*c,d*). The group of nerve cells at the base of each arm is the site of coordination, and any one of the five nerve cell groups can control the direction of tube feet stepping in the animal as a whole. In this way a starfish can make a temporary brain without having a permanently defined brain structure.

17.3.4 Ophiuroidea (brittle stars)

The Ophiuroidea constitute another large class (2000 species) of rather smaller starfish. They have a very much harder skeleton than do asteroids and, although they are stellate in form, the arms are clearly marked off from the disc (Figure 17.2*e*). Most of the arm consists of central muscle surrounded by ossicles (Figure 17.2*f*) which are able to move on each other and give the arms great flexibility (hence the name 'serpent stars'). These ossicles are often called 'vertebrae' and indeed no other invertebrates have such vertebrate-like struc-

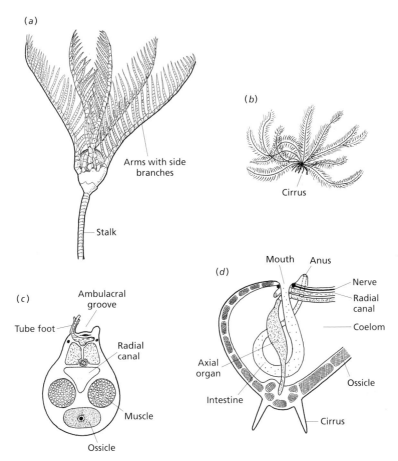

(a)

Arms with side
branches

Stalk

(b)

Cirrus

(c)

Ambulacral
groove

Tube foot

Radial
canal

Muscle

Ossicle

Mouth   Anus

(d)

Nerve

Radial
canal

Coelom

Axial
organ

Intestine

Ossicle

Cirrus

Figure 17.2.   The anatomy of crinoids and ophiuroids: (a) *Ptilocrinus pinnatus*, a sessile crinoid; (b) *Antedon*, a feather star, a motile crinoid; (c) transverse section (T.S.) of arm, *Antedon*; (d) vertical section (V.S.) of crinoid disc;

tures. Most ophiuroids resemble asteroids in being active carnivores, but their methods of feeding and locomotion are very different. Food is seized by the mobile arms, assisted by mucus secretion from tube feet that have no ampullae and play only a minor role. Ophiuroids walk by two arms on each side sweeping back from the leading arm; this 'rows' the animal along in a series of jerks, with the spines providing traction against the ground. Swimming is achieved by similar arm movements. Coordination of these relatively rapid movements is more highly developed than in asteroids. The arms not only have 'segmental' ossicles, they have a corresponding series of nerve ganglia along each arm. Fibres termed 'giant' in contrast to other echinoderm nerves

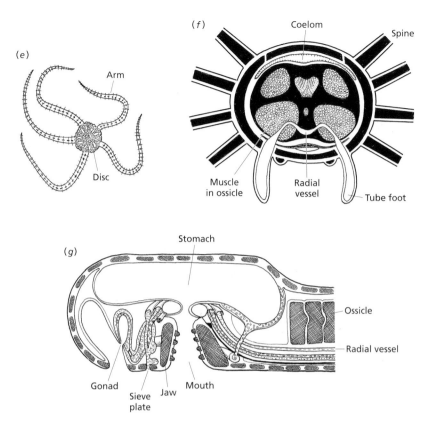

Figure 17.2 (*cont.*). (e) dorsal view of a brittle star (ophiuroid); (f) T.S. arm, *Ophiothrix* with extensive ossicle (shown black); (g) V.S. ophiuroid.

are large enough to be investigated with electrodes and their structure and physiology is known. Dominance by the leading arm appears to depend on the action of the arm's nervous tissue as a whole.

17.3.5 Echinoidea

The Echinoidea (some 950 species) include 'regular' echinoids (sea urchins, Figure 17.4*a,b*) and 'irregular' ones (heart urchins, sand dollars and others, Figure 17.4*c*). Unlike starfish and brittle stars, sea urchins are slow-moving browsing herbivores. They are spherical or egg-shaped, with the axis between the oral and aboral surfaces much extended. Uniquely among echinoderms, the body is encased in a fixed lattice of ossicles, which may be fused to form

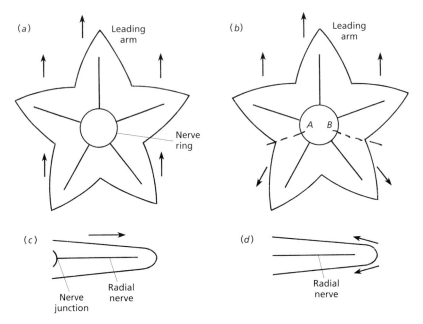

Figure 17.3. Experimental demonstration of the coordination of tube feet by nerves in a starfish. Arrows denote the direction of stepping of tube feet. (a) Intact starfish; (b) with the nerve ring cut at A and B; (c) isolated arm, including the junction between nerve ring and radial nerve; (d) isolated arm, without the junction.

a rigid 'test'. In place of arms there are five ambulacral grooves, where tube feet emerge between small skeletal plates (see Figure 17.4b). The body is covered in spines, and pedicellariae (pincers) are important in removing settling organisms.

Sea urchins feed by scraping seaweed from rocks with an elaborate structure of movable skeletal plates called 'Aristotle's Lantern'. Locomotion is mainly by spines that articulate with the main skeleton by ball and socket joints (Figure 17.4d) and are controlled partly by muscle and partly by mutable connective tissue. The tube feet are very well developed, with ampullae and suckers. They assist locomotion and keep the urchins close to the substratum on which they are grazing or anchor them in crevices.

Irregular echinoids are deposit feeders adapted to burrowing shallowly under sand. Some are greatly flattened and covered in small spines, with bilateral symmetry superimposed on a radial plan. The tube feet are very diverse in structure and function, some being modified for digging and others as respiratory 'siphons' leading up to the surface of the substratum.

17.3.6 Holothuroidea (sea cucumbers)

This comprises 900 species which are very unlike other echinoderms. They are secondarily bilaterally symmetrical, lying on one side with an elongated body axis between the mouth and the anus (Figure 17.5*a,b*). The endoskeleton is very much reduced, leaving a muscular body wall with a few embedded ossicles but without spines or pedicellariae. There is only one gonad. Respiratory trees, branches of the gut taking in oxygenated sea water, are characteristic.

Typically holothuroids are deposit feeders. A group of anterior tube feet forms long buccal tentacles that are held out as a net or gather food from the sand. Locomotion is slow, by worm-like wriggling of the muscular body wall, or there may be locomotor tube feet in ambulacral grooves on the underside of the body. When pursued by predators, holothuroids may immobilise them by extruding a mass of blind-ending tubules, which are very sticky, or may break off the internal organs and leave them in the path of the predator: the holothuroid escapes and regenerates its viscera.

17.3.7 Concentricycloidea (sea daisies)

The Concentricycloidea were discovered only in 1986 in deep seas off New Zealand. Two species of *Xyloplax* have been named. They are at present considered to constitute a separate Class fairly closely related to Asteroidea. Superficially they resemble medusae (Figure 17.5*c*) but there is a water vascular system with two connected concentric rings, one of which bears tube feet. The spermatozoa are very unusual and, uniquely to the phylum, fertilisation may be internal. Feeding methods are unknown (one of the species has no gut) but absorption of dissolved organic materials is probably important.

17.4 Do the larvae illuminate echinoderm evolution?

Most echinoderms have motile planktonic larvae. These are believed to be primitive to the phylum: and indeed some species that develop directly from yolky eggs contain vestigial larval structures. Planktonic larvae are an advantage to sessile or slow-moving animals living on the sea bottom, because they can exploit a different food source and can disperse the species. They also allow greater fecundity, since a large number of small eggs without food reserves can be produced.

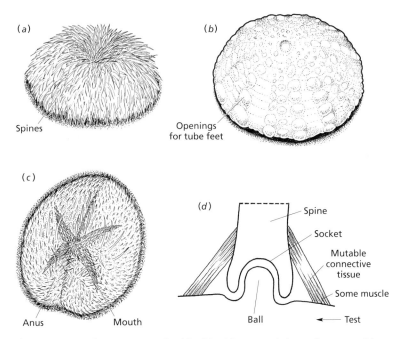

Figure 17.4. The anatomy of echinoids: (a) external view of a sea urchin (echinoid); (b) a sea urchin 'test' (skeleton); (c) a sand dollar, an irregular echinoid; (d) the mobile spine of a sea urchin.

Figure 17.6 indicates the larvae characteristic of the different Classes. The prevalence of convergent evolution is instantly apparent. Larvae are, above all, adapted to their own way of life (Box 17.1). That both ophiuroids and echinoids have a 'pluteus' larva with the ciliary band extended into arms by eight skeletal rods does not mean that brittle stars and sea urchins have an improbably close relationship, but that both Classes have the same adaptation, which extends their surface area thereby both facilitating floating by increasing the viscous drag and increasing the feeding surface (some echinoid larvae grow longer arms when food is short). Some ophiuroids have larvae similar to the crinoid doliolaria, derived by loss of the arm rods, but this is no link between ophiuroids and crinoids. Nor has the resemblance between the asteroid bipinnaria and the holothuroid auricularia any phylogenetic significance except that both are developments of the basic 'dipleurula' larva (Figure 17.6a) in which form many echinoderms hatch. All these larvae are bilaterally symmetrical; the adult rudiment is set aside in the late larva and radial symmetry first appears when it develops.

Convergent evolution extends much further than the external form of different larvae. The loss of planktonic larvae from the life cycle has occurred

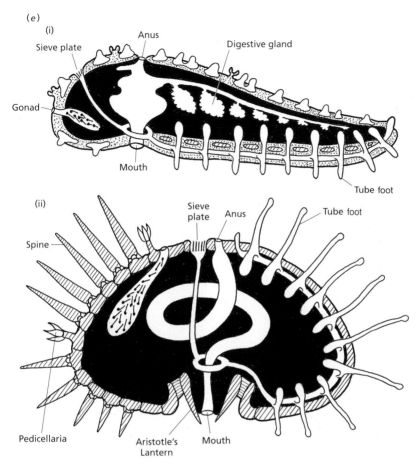

Figure 17.4 (*cont.*). (e) Diagram comparing the general form in vertical section of (i) a starfish and (ii) a sea urchin.

many times in unrelated groups, and the same series of steps can be traced: first the eggs become larger, then the larvae lose the ability to feed and then the time before metamorphosis is reduced. A feeding larva has been lost in all present-day crinoids, many ophiuroids, and 14 times in five different Orders of echinoids. Total loss of larvae has occurred in several ophiuroids and in at least six echinoid lineages. These changes may or may not be accompanied by smaller adult size, viviparity and brooding behaviour. In asteroids also, simplification and reduction of larval form has occurred many times separately, as has the change to hermaphroditism or to viviparity.

Larval morphology clearly has diversified independently of adult morphology, and convergent adaptations are rife. Can larval structure illuminate phylogenetic relationships at all? Probably only in those characters that are the

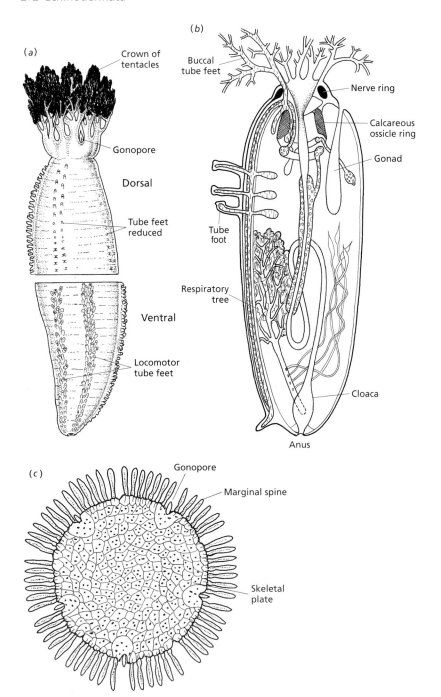

Figure 17.5. The anatomy of holothuroids and concentricycloids: (a) diagram of *Cucumaria frondosa*, a holothuroid, (upper part viewed dorsally, lower part ventrally); (b) vertical section of holothuroid; (c) dorsal view of *Xyloplax medusiformes*, a concentricycloid.

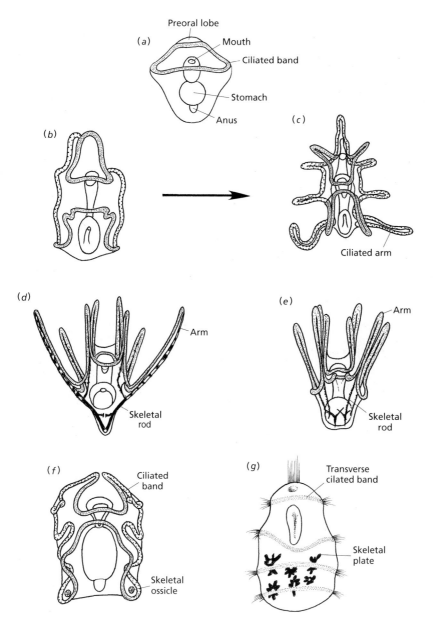

Figure 17.6. Echinoderm larvae: (a) dipleura, the first larva of most groups, showing bilaterally symmetrical form; (b) bipinnaria, asteroids; (c) brachiolaria, asteroids; (d) ophiopluteus, ophiuroids; (e) echinopluteus, echinoids; (f) auricularia, holothuroids; (g) doliolaria, crinoids.

Box 17.1
Larvae

What is a larva?

A larva is a pre-adult form in which some animals hatch from the egg. Further definition runs into difficulties: the larva is different in form from the adult, but the difference may be confined to smaller size and immaturity, as in the nematodes and exopterygote insects, where the series of larvae simply represent a discontinuity imposed on growth by the need to moult. A larva is sexually immature, but see the discussion of paedomorphosis in Chapter 18. Characteristically it is a feeding stage in the life cycle, but many larvae do not feed (e.g. the short-lived cypris larva of barnacles and the tunicate tadpoles, both of which serve simply to disperse the sessile adults). Larvae often are dispersal agents (as in many crustaceans) but not for example in insects, where the adult may have wings.

Why have a larva?

Early hatching can be an economy if it saves the adult from providing large quantities of yolk. Larger numbers of young can be produced and the reproductive turnover will be faster, but most larvae are fragile and very many will be eaten or die. Larvae often exploit a food source different from the adult's and may colonise a different habitat, thus avoiding competition between adult and young. They may both disperse the species and find a suitable habitat for the adult.

Which invertebrates have larvae?

All phyla have examples of life histories including larvae (see figures in the relevant chapters).

Marine invertebrates

The marine plankton is rich in food, readily disperses animals, and small floating forms do not require special adaptations. In annelids and crustaceans, for example, marine animals nearly all have larvae while freshwater and terrestrial animals do not.

Sessile animals

Slow-moving, bottom-living animals almost always use larvae for dispersal. Examples include sponges, hydroids, anemones and corals, marine molluscs other than cephalopods, barnacles and echinoderms.

Box 17.1 *(continued)*

Parasites

Larvae are important agents of transfer between parasite hosts. Often there is an elaborate life cycle with a succession of hosts, commonly with one preying upon another, and the volume of reproduction may be increased by asexual reproduction of some larval stages. (See Figure 6.4a,b, platyhelminth life cycles).

Arthropods

Arthropods need to moult as they grow and therefore readily incorporate metamorphosis in the life cycle. Larvae can be very unlike the adults.

What can larvae tell us about phylogeny?

The answer has to be 'very little'. Larvae are adapted to their environment. However, larvae are the products of early hatching, and therefore may show in their structure the consequences of earlier embryonic development, such as the method of cleavage and of mesoderm formation, which do have phylogenetic significance. Many protostomes, for example, hatch as larvae rather like trochophores: this does not necessarily relate them closely to annelids or to molluscs, but it does distinguish them from echinoderms and chordates. Occasionally aberrant forms (such as *Sacculina*, see Chapter 13) can be placed by larval resemblances, but more often such resemblances can be attributed to convergence.

This brief survey of larvae is very far from complete, even concerning invertebrates mentioned in this book (e.g. see the sessile scyphistoma larva of many jellyfish (Chapter 4) and the great range of aquatic and terrestrial larvae of insects (Chapter 15)).

---

immediate consequences of early development being deuterostome. For example the deuterostome mouth (see Box 5.3) is surrounded by cilia, which become the single feeding band of the earliest larva retaining food particles on an upstream beat. This is significantly different from the trochophore larva of protostomes, where the ciliary bands are differently arranged and trap food particles as they travel downstream. The larvae of other deuterostomes, such as hemichordates and protochordates, capture their food in the same way as echinoderm larvae.

The conclusion is that we can learn about the process of evolution from echinoderm larvae (the selective advantage of having a larval stage in the life cycle and the adaptations which have occurred) but, owing to the abundance of convergence in larval forms, we can learn very little about the course of evolution. Chapter 20 explains how we can trace the course of echinoderm evolutionary history from the particularly good fossil record for many classes of echinoderms, and molecular methods (including analysis of the genes controlling early development) can help us further.

18

Invertebrate Chordata

The phylum Chordata consists mainly of the Class Vertebrata (animals with backbones, i.e. the fish, amphibians, reptiles, birds and mammals including ourselves). There are, however, some smaller groups of animals that do not have backbones but possess the distinctive chordate characters. All are marine, and like the echinoderms they are deuterostomes. These animals are interesting in that they demonstrate another variety of invertebrate life, and provide useful information about the origin of vertebrates.

18.1 What are the chordate characters?

Figure 18.1 shows the notochord, the hollow dorsal nerve cord, the pharyngeal gill slits and the postanal tail. All these characters are related to swimming by lateral undulations of a body like a tadpole or a fish.

1. **The notochord** gives the phylum its name (avoid verbal confusion with 'nerve cord'). The notochord is a dorsal stiffening rod running from near to the anterior end to the posterior tip of the body. It is made of close-fitting cells with hydrostatic pressure due to intracellular vacuoles, enclosed in a fibrous sheath; it arises from the roof of the embryonic gut. It is longitudinally incompressible but laterally flexible, and serves as an energy-saving device, because any animal swimming by muscular undulation must be stiffened to localise the effects of muscle contraction. A swimming leech, for example, uses energy in contracting strong dorsoventral muscles while it generates the muscular wave that pushes it

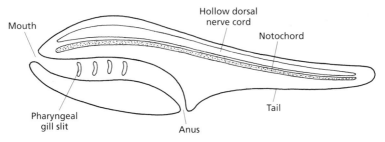

Figure 18.1. Diagram to show chordate characters.

through the water, while a chordate is sufficiently stiffened by the notochord.

2. **The hollow dorsal nerve cord**, immediately dorsal to the notochord, innervates the main swimming muscles of chordates that lack legs. In many other animals, notably annelids and arthropods, the main nerve cord is solid and ventral in position. In chordates the principal blood vessels are ventral and the blood flows forward ventrally, unlike annelids and arthropods where the blood flows forward dorsally. These differences long ago prompted the suggestion that chordates evolved from 'worms turned upside down'; recent genetic analysis has revived controversy about this suggestion (Chapter 20).

3. **Pharyngeal gill slits** primitively serve filter-feeding, from a stream of water due either to the animal swimming through the sea or to ciliary beating bringing sea water through the animal. Water enters by the mouth, folds of skin (gills) aided by mucus trap the food particles, and the water leaves the pharynx through the gill slits instead of passing right through the gut to the anus. Chordates feeding on larger masses of food may use the gills for respiratory exchange and retain the gill slits to let water out.

4. **The postanal tail**, like the notochord, is concerned with efficient locomotion, being primitively a muscle-packed region exerting propulsive force.

These characters may be lost or modified in many members of the phylum, for example in most vertebrates the embryonic notochord is replaced during development by the vertebral column (backbone) and in ourselves the gill slits are only briefly marked in the early embryo and the tail is vestigial. Many invertebrate chordates show the chordate characters only at certain stages of the life cycle.

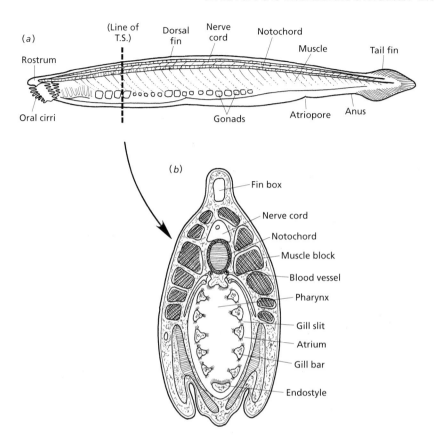

Figure 18.2. Cephalochordates: (*a*) longitudinal section of the amphioxus (*Branchiostoma*); (*b*) transverse section (T.S.) of the amphioxus (*Branchiostoma*).

18.2 Which are the invertebrate chordates?

18.2.1 Cephalochordata

The Cephalochordata (25 spp., up to 10 cm long) show the chordate characters very clearly. The lancelet or 'amphioxus', *Branchiostoma,* is the main example (Figure 18.2). The amphioxus tapers to a point at both ends; the simple epidermis extends into dorsal, caudal and ventral fins, and by transparency the V-shaped body muscles are very obvious from outside. It can swim by lateral undulations but spends most of its time half buried in shallow marine gravel: it can still be found by dredging near the Eddystone lighthouse in Plymouth Sound, England. The amphioxus is specialised in that the muscles and their nerves are asymmetrical in arrangement and the gills are greatly elaborated and multiplied, with accessory ciliary tracts and an external oral wheel. Food particles are trapped in mucus secreted by a structure called the endostyle on the floor of the pharynx. The gill slits are very numerous and do not open directly to the outside world (so many perforations would greatly weaken the body wall) but into an atrium, which opens more posteriorly by an atriopore. The atrium is formed by folds of the body wall in the pharyngeal region, which appear in the late larva and grow down to the ventral surface, where they fuse (Figure 18.2a,b). The notochord is unique in extending right to the anterior tip of the body. The anterior end of the dorsal nerve cord is not expanded into a brain, nor is there a heart; the blood is circulated by general contractility of the blood vessels. The excretory organs are diffuse and not well developed, but they are not the worm-like protonephridia of erroneous earlier descriptions. The sexes are separate, fertilisation is external and the zygote hatches early as a small free-swimming version of the adult.

18.2.2 Urochordata (Tunicata)

These are our most improbable cousins, revealed as such only by their larvae. This 'tadpole larva' (Figure 18.3a) shows all the chordate characters, but at metamorphosis it turns into a sea squirt or a colony of similar organisms, enclosed in a gelatinous tunic.

Ascidiacea

Ascidians (about 2000 spp. Length 0.1 cm to 1 metre) are sessile filter feeding 'sea squirts', common in shallow sea water. There is no head and the nervous system is poorly developed. The gut is U-shaped between two openings, the buccal and atrial siphons, and the much enlarged pharynx is perforated by many gill slits; this is the only chordate character remaining in the adult (Figure 18.3b). As in the amphioxus, ciliary feeding is assisted by mucus secreted by an endostyle and the gill slits open into an atrium. Unique features of ascidians include a heart that periodically reverses its beat and the enclosure of the whole body in a gelatinous tunic that is strengthened by cellulose fibres and penetrated by blood vessels. The tunic is not moulted but grows with the animal. Another unique feature is that, as sea water is passed through the animal, vanadium ions are retained and concentrated, mainly in the blood cells. Apparently the vanadium ions assist the polymerisation of fibres in the tunic, and may also act as antibiotics.

Ascidians are nearly all hermaphrodites, usually cross-fertilised, and the zygote hatches as the tadpole larva, which is primarily an agent of dispersal. Great powers of regeneration and budding are characteristic of ascidians, as is manifested by compound ascidians. Individuals are combined in a series of star-shaped colonies, very common on the seashore and often brightly coloured, embedded in jelly. Each colony has separate mouths at the points of the star and a common atrial opening in the centre (Figure 18.3c).

Thaliacea

The Thaliacea (70 spp.) are floating transparent tunicate colonies (Figure 18.3d), with buccal and atrial sinuses at opposite ends of the colony. The water current, generated by muscular contraction of the body wall, is used for swimming by jet propulsion as well as for filter feeding and respiration.

Larvacea

The Larvacea (70 spp.) are best known from *Oikopleura* (Figure 18.3e). They are more obviously chordate than other adult tunicates, since the larval form is retained following accelerated early development. The adult floats in the plankton enclosed in a gelatinous 'house' (without cellulose fibres) through which tail movements propel a stream of water. There is only one pair of gill slits, and the food consists of very small food particles trapped in mucus. The house is shed and reassembled at frequent intervals.

Sorberacea

These deep water forms comprise a few species of most remarkable carnivorous tunicates. Prehensile finger-like extensions of the buccal siphons stretch out and capture small worm and crustacean prey.

18.2.3 Hemichordata

The Hemichordata have the most problematic relationships of all the invertebrate chordates. They are sometimes included with Cephalochordata and Urochordata as 'Protochordata' but are more often classified separately, always with disclaimers expressing uncertainty. They consist of two groups very dissimilar in appearance.

Enteropneusta

The Enteropneusta (70 spp., length 9–200 cm) are large solitary worm-like animals such as *Saccoglossus* (= *Balanoglossus*), burrowing in the mud or sand of the shallow seas. The body consists of three regions, a proboscis, collar and trunk (Figure 18.4a,b), corresponding to the three divisions of the embryonic

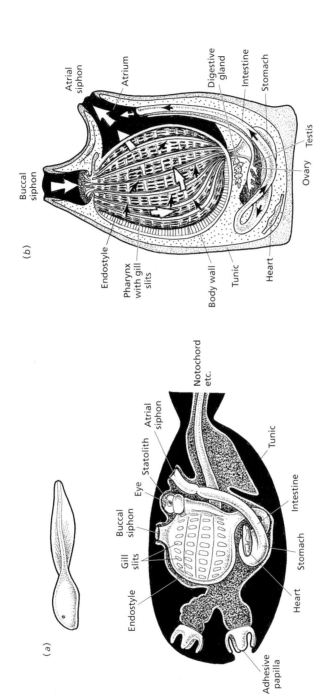

(a)

Notochord
etc.

Eye   Statolith   Atrial
                  siphon

Buccal
siphon

Gill
slits

Tunic

Endostyle

Intestine

Stomach

Heart

Adhesive
papilla

(b)

Atrial
siphon

Atrium

Digestive
gland

Intestine

Stomach

Buccal
siphon

Testis

Endostyle

Ovary

Pharynx
with gill
slits

Body wall

Tunic

Heart

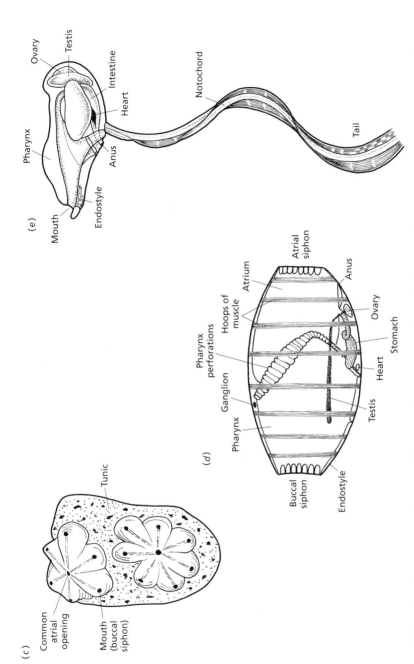

Figure 18.3. Urochordates: (a) diagram of the tadpole larva of a tunicate; (b) diagram of vertical section of *Ciona*, an ascidian; (c) *Botryllus*, a colonial ascidian; (d) *Salpa*, a thaliacean; (e) *Oikopleura*, a larvacean.

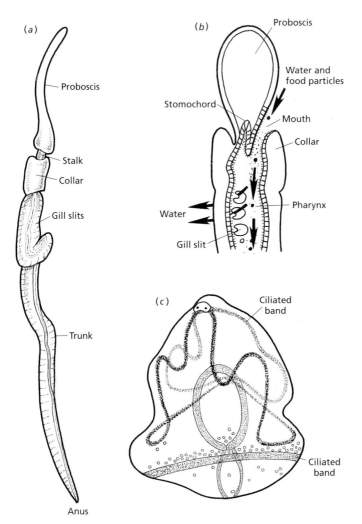

Figure 18.4. Hemichordates: (*a*) *Balanoglossus*, a solitary enteropneust; (*b*) diagrammatic longitudinal section of the anterior end of an enteropneust; (*c*) tornaria larva of an enteropneust;

coelom. The ciliated proboscis collects food into the mouth, which opens from the collar. The trunk is perforated by a long row of pharyngeal gill slits, which serve primarily for respiration rather than filter feeding; there is no mucus-secreting endostyle.

Gill slits are the only chordate character found in hemichordates. The larva is not a tadpole but a 'tornaria', remarkably similar to the auricularia of holothuroid echinoderms (Figure 18.4*c*, compare Figure 17.6*f*).

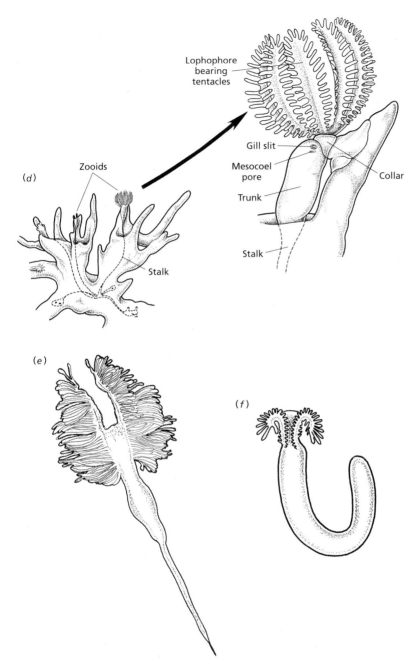

Figure 18.4 (*cont.*). (*d*) *Cephalodiscus*, colony with one zooid enlarged; (*e*) a fossil graptolite; (*f*) a lophoenteropneust, from a deep sea photograph.

Pterobranchiata

The Pterobranchiata (21 spp. in two genera, *Cephalodiscus* and *Rhabdopleura*) are minute (1–5mm) sessile and colonial animals with lophophores (Figure 18.4*d* and Figure 16.2a). The zooids are ciliated all over and collect food externally. *Cephalodiscus* has a single pair of gill slits; *Rhabdopleura* has none. They reproduce asexually by budding or sexually by releasing gametes. Early development is unambiguously deuterostome, but the larva, unlike the enteropneust tornaria, is uniformly ciliated and short lived. It has a store of yolk and does not feed, serving solely for dispersal.

Pterobranchs can be traced back in the fossil record. Fossil graptolites (Figure 18.4*e*) were known from the Cambrian to the early Devonian. They were once considered to be cnidarians, but further study diagnosed them as pterobranchs. There was great excitement when in 1989 a living graptolite was hauled up from the deep sea and could be assigned to a modern genus, as *Cephalodiscus graptolitoides*.

Why are enteropneusts and pterobranchs combined in the same phylum? Two unique characters unite them, the stomochord and the collar. The stomochord is a rod-like hollow process growing out from the gut dorsally, lined with flagellated cells. It projects forward from the roof of the mouth into the proboscis, where it supports the heart and excretory organs, in enteropneusts, or to the cephalic shield, in pterobranchs (see Figures 18.4*b,d*). This rod looks like a short notochord, but is not homologous with it. The collar supports the proboscis and contains the main concentration of nervous tissue, which is formed by invagination (as in all chordate embryos) and entirely epidermal in origin and position, being rather similar to the echinoderm nervous system. The collar contains the middle division of the coelom, the mesocoel, opening by mesocoelic ducts either into the first gill slit (enteropneusts) or just beside it (in the pterobranch *Cephalodiscus*). Molecular evidence (from 18SrDNA) also suggests that the two groups should be placed in the same phylum.

'Lophoenteropneusta', 1 metre long on the deep sea floor, have a proboscis that secretes mucus, trapping food particles. This proboscis is flanked by lophophores (Figure 18.4*f*). These forms bear living witness to the relationship in that they have the general form of enteropneusts combined with the lophophores of pterobranchs.

18.3 What are the relationships of invertebrate chordates?

That the **cephalochordates** are closely related to craniates (vertebrates) is firmly established. Morphology and embryonic development unite these groups beyond reasonable doubt, and the only question has been whether the amphioxus is primitively simple or has secondarily lost the vertebrate specialisations. Study of *Hox* genes (see Chapter 19) confirms both the closeness of the relationship and the primitive nature of the amphioxus. **Urochordates** are very much harder to place. A classic early account by W. Garstang, followed by N. J. Berrill, invoked 'neoteny', where the larval form becomes sexually mature; if a tadpole larva were to reproduce, the result would be an animal very like the amphioxus (or at least, very like its larva before adult specialisations develop). The sessile and gelatinous tunicate adult would then be entirely omitted from the life cycle. Species showing both kinds of life cycle (with an adult in the sessile tunicate or the chordate form) survive to the present day, placed in different phyla.

The phenomenon of 'paedomorphosis' (reproduction in the larval form) is known from a number of present-day animals. It is called 'progenesis' when gonadal development is accelerated and reproduction occurs earlier in the life cycle, or 'neoteny' when somatic development of adult tissues is delayed and reproduction occurs at the time normal to that species but while it is still a larva. Paedomorphosis is one way in which the normal process of natural selection can result in sudden abrupt changes in form. It might have occurred several times in the chordate lineages: larvaceans such as *Oikopleura* might be products of neoteny within the tunicates, and the vertebrates resemble the amphioxus larva rather than the specialised adult.

The theory of group origins by neoteny is attractive but hard to prove, and molecular evidence does not seem to support it. Preliminary 18SrDNA results suggest that larvaceans diverged early from the main stock, and that adult tunicates represent a secondary specialisation rather than an early adult form that has been discarded. At the present time, while it is clear that the urochordates, cephalochordates and vertebrates are closely related, it is not possible to trace the course of evolution with certainty. What is certain is that the chordate characters, at whatever point they arose, proved to be excellent adaptations to a free-swimming life in the sea.

Where do the **hemichordates** fit in, and should they be included as chordates at all? Both pterobranchs and enteropneusts collect their food externally, unlike other early chordates. Pterobranch lophophores are remarkably similar to those of the protostome lophophorate phyla, and external food collection occurs in enteropneusts even though they have pharyngeal gill slits and the structure of the gill skeleton resembles that of the amphioxus. Larval structures do not provide compelling evidence, although the dorsal structures of a chordate might be derived from the ciliary bands of a tornaria.

At this point a wider view of deuterostome relationships is required, to include the echinoderms, the only other certain deuterostomes. Hemichordates and echinoderms develop very similarly, not only in the earliest stages that define deuterostome embryology but also in the origin and development of the three divisions of the coelom. For example, the water vascular system in echinoderms is derived from the middle coelomic pouch, as is the collar coelom, which penetrates throughout the lophophore in pterobranchs. The sessile crinoids collect their food using tube feet containing a branch of the water vascular system, very much as the sessile pterobranchs use the corresponding structure, the lophophore. The echinoderm madreporite corresponds to the hemichordate collar pore. In both phyla the left coelomic pouches become dominant as the animal develops, and the pouches on the right side are lost. Such morphological resemblances are unlikely to be due to convergence, since molecular evidence (from 18S rDNA) also suggests a close relationship between echinoderms and hemichordates.

Once again, we are faced with an unresolved trichotomy. Echinoderms, hemichordates and the rest of the chordates appear to be related, but we do not know exactly how; that is, we cannot trace the course of evolution among them. There is some hope of help from the study of fossil echinoderms, but the interpretation of some of the main fossils is controversial.

19

Development

19.1 How do animals develop?

Development of an individual organism (ontogeny) begins with a single cell, most commonly a fertilised egg. To make a multicellular organism this cell must repeatedly divide, and the daughter cells must grow and become different from each other. Cells must move as they become organised into a body. Development depends upon the coordinated behaviour of cells, controlled by information supplied by genes, and must produce the right sort of animal every time.

– **Cell division**: cleavage, the earliest divisions of the fertilised egg, proceeds by DNA replication and subdivision of the cytoplasm, normally without any increase in size (growth occurs later).
– **Cell differentiation** is preceded by **commitment** (often called 'determination'). This is the limitation of developmental capacity, directing the cell to a particular pathway. This may happen long before there are any visible signs of differences between cells, and the timing of commitment in relation to cleavage varies in different kinds of animal.
– **Gastrulation** follows cleavage. It is a series of **cell movements** that bring the cells that will form internal organs to the inside of the developing embryo.
– **Pattern formation**: the organism has to be given a pattern as a whole, and the first step is the establishment of axial polarity and the interactions of cells that give them positional information.
– **Growth**: at each stage in the laying down of the basic pattern of the

organism, there can be some growth. Cells increase in size and multiply, and intercellular materials (including fluids and hard skeletons) are produced.

This chapter first introduces these main processes of early development and their sequence and detailed mechanism in different invertebrates, using four main examples: the gastropod mollusc *Lymnaea*, the sea urchin *Echinus*, the nematode *Caenorhabditis* and the insect *Drosophila*. Our present understanding of the genetic control of development is then introduced.

19.2 What makes different animals develop differently?

Development is directly or indirectly under gene control at every step, and within most species the developmental process varies little between individuals. Environmental factors can, however, modify development; for example, temperature may determine the rate and even sometimes the kind of development (as in those animals whose sex is determined by ambient temperature), and the polymorphism between females within a species of social insects may be environmentally determined. Interspecific developmental differences may be associated with the reproductive strategies of different species, rather than with phylogeny (Figures 19.1 to 19.4). The quantity of yolk that the eggs contain is particularly important. Large quantities of yolk feed the embryo for longer, allowing longer protection within the egg and later hatching, but yolk is heavy and inert, hampering the division and the movement of cells. If there is very little yolk, many more eggs can be produced, the embryo can divide faster and gastrulate unimpeded, but then it needs food. It may be fed by the mother within her body, but while many invertebrates retain and protect developing eggs ('ovoviviparity'), few have true viviparity, where the embryo is fed through a placenta, as in mammals and some reptiles. Viviparity occurs in *Peripatus*, which feeds its young on internally secreted 'milk' and some insects (for example, tsetse flies), but most invertebrate eggs with little yolk hatch very early into a form that can find its own food.

19.3 What is the pattern of cleavage in invertebrates?

Cleavage is the division of the fertilised egg to form a mass of cells, called 'blastomeres'. All the products of cell division in the embryo normally contain identical sets of genes, which do not change in the course of development (see Figure 19.5 for an experiment elegantly demonstrating this in a tadpole). In any cell only some of the genes will be expressed, which shows that gene expression must be regulated. Often the cytoplasm is the initial source of cell diversity, since it contains substances localised in different regions of the egg. These substances may be proteins and long-lasting messenger RNA supplied by the mother. The plane of successive cleavages is constant for a given group of animals: information contained in the cytoplasm is therefore predictably partitioned between the daughter cells by the process of cleavage. In short, it is at first the mother's genes that control the expression of the embryo's genes and the mechanics of cell division.

The pattern of early cleavage may reveal phylogenetic affinity. In animals with three cell layers it largely reflects a major division (see Box 5.3). Spiral cleavage is characteristic of protostomes. It probably evolved once only, so that minor differences in the mechanics in different groups may be significant indicators of phylogeny. After the four-cell stage, each daughter cell is set vertically above the furrow between the two cells beneath it, forming a spiral as cell numbers increase (Figure 19.1). In *Lymnaea*, whether the cleavage spiral coils to left or to right is determined at the stage when only maternal genes are acting. Later the shell coils in the same direction as the cleavage spiral. (This explains a long-standing puzzle as to why the direction of coiling of the shell depends solely on the genes in the mother.) The first steps of commitment occur at some stage in cleavage. Spirally cleaving eggs were formerly called 'mosaic' eggs because, in an extreme example, separating the first two blastomeres may result in each developing half an embryo. This is evidence for early commitment. In these embryos the mesoderm is set aside very early: it can be traced back to the blastomere, which, in a formal system of labelling the spiral, is called '4d', a division product of D (see Figure 19.1). If the 4d cell is removed, a deformed trochophore may develop, lacking all mesoderm.

Radial cleavage, where daughter cells are set immediately above the cells below (Figure 19.2), occurs in the deuterostome phyla (echinoderms and chordates) and also in phoronids, bryozoans, brachiopods, chaetognaths, rotifers, priapulids and a few others. Experimental work on sea urchins in the early twentieth century showed that separation of the first two blastomeres results in regulation of each to form a small complete embryo: this result led to slightly misleading descriptions of radially cleaving eggs as 'regulative'. Certainly, full commitment does not occur until later in cleavage. Future mesoderm also is not defined early, it arises later by outpouching from the gut (see Figure 19.2).

There are other cleavage patterns. Among both sponges and cnidarians there are many different cleavage patterns. The ctenophore plate and the

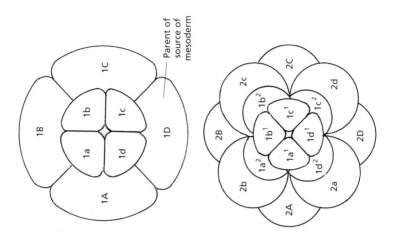

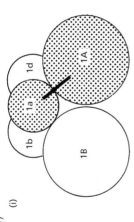

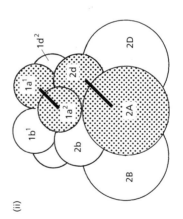

(a)

(i)

(ii)

1d

1a

1b

1A

1B

Parent of
source of
mesoderm

1b   1b

1a   1d

1B   1C

1c

1D

1A

2c   2C

2B   1b²   1b²   2d

1b¹   1c¹   1d¹

2A   1a¹   1c²   2D

1a²   1a¹   1d¹

2b   1d²   2a

2B   2A   2D

1d²

1a¹

1a²   2d

1b¹   2A

2b

2B   2D

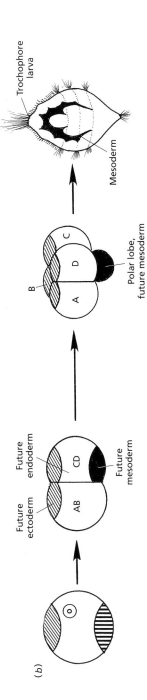

Figure 19.1. Molluscan early development: (a) spiral cleavage, for example *Lymnaea*, viewed laterally (left hand column, with cell 1A and its derivatives shaded) and dorsally (right hand column) at the stages of (i) 8 cells and (ii) 16 cells; (b) mesoderm localisation in *Dentalium*.

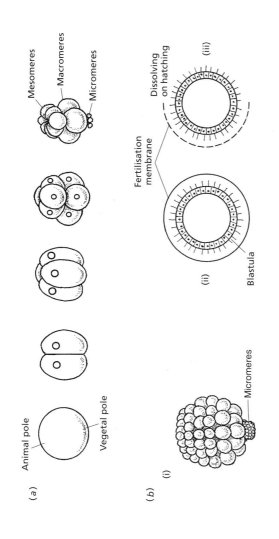

(a)

Animal pole

Vegetal pole

Mesomeres

Macromeres

Micromeres

(b)

(i)

Micromeres

(ii)

Blastula

Fertilisation
membrane

(iii)

Dissolving
on hatching

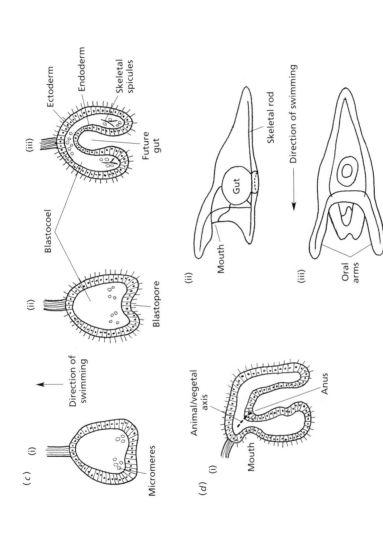

Figure 19.2. Echinoderm (sea urchin, e.g. *Echinus*) early development, with the animal pole uppermost in all diagrams: (a) successive stages of radial cleavage; (b) blastula, 64 cell stage, (i) lateral view, (ii) section of blastula within fertilisation membrane and (iii) hatching; (c) gastrulation (i) about to start, (ii) when blastopore appears and (iii) when future gut appears; (d) larva formation, (i) diagram of early larva, (ii) pluteus larva in side view and (iii) pluteus larva in ventral view.

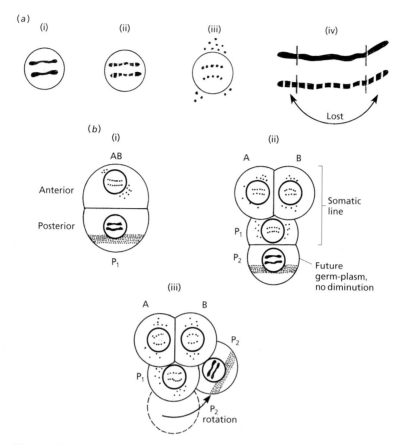

Figure 19.3. Nematode (*Ascaris*) early development. (*a*) Chromosome diminution: (i) two large chromosomes, (ii) to (iv) fragmentation and loss of one chromosome. (*b*) The first stages of cleavage: (i) separation of AB from P_1; (ii) the four-cell stage with the nematode T; (iii) the final form of the four-cell stage.

acoelan bispiral (mentioned in Chapters 5 and 6) are unusual arrangements of the first blastomeres, and the nematode **T** is shown in Figure 19.3. Whatever the pattern of the first divisions, the end-product of cleavage in these embryos is a ball of cells, the blastula, which is hollow because the cells secrete fluid into the centre.

Most arthropods and many other invertebrates have large quantities of yolk, impeding cleavage. Some invertebrates (e.g. some cnidarians) leave the yolk undivided outside the embryonic cells, a condition familiar from the eggs of birds, where the blastula is a flattened disc lying on top of a mass of yolk.

Most arthropod eggs, however, have a large quantity of central yolk, with the embryonic cells peripheral to it. In insects the fertilised egg nucleus divides rapidly many times to form a syncitium, each daughter nucleus having a small surrounding mass of cytoplasm with no cell boundary. The nuclei with adhering cytoplasm migrate from the centre to the periphery, where cell boundaries form (Figure 19.4).

19.4 How do invertebrates gastrulate?

The cell movements constituting gastrulation vary considerably among sponges and cnidarians, but the process is very similar in most animals with three cell layers. Most commonly, cells move from the surface of the blastula to the interior by 'invagination', the infolding of a sheet of cells (see Figure 19.2). At a point on the blastula surface called the blastopore some cells move inward, largely obliterating the lumen of the blastula and making an internal layer, the endoderm, which encloses the future gut cavity (see Box 5.3). The blastopore becomes the mouth in molluscs and nematodes, the anus in echinoderms. Cleavage of very yolky eggs produces a plate of cells, not a ball, but the invagination process is very similar; for example, in insects gastrulation occurs within a thickened band of the superficial cells (see Figure 19.4). An alternative method called 'epiboly' occurs, for example, in some platyhelminths, where a sheet of outer cells spreads around an inner layer, which hollows out to form the gut.

The cell movements constituting gastrulation affect the entire embryo. The primary germ layers are defined; changes in surface cell adhesion molecules enable cells of the same layer to recognise each other and to stick together. In triploblastic animals the outer layer (ectoderm) gives rise to the surface epithelia and the nervous tissue, the innermost layer (endoderm) forms the gut and its glands and the mesoderm forms muscles, gonads and everything else in between the ectoderm and endoderm. During gastrulation a pattern of future cell identities previously laid down in two dimensions is transformed into a three-dimensional embryo.

19.5 How is polarity established and maintained?

This question takes us back to an earlier stage in development. Embryonic development involves a number of cell activities, organised in place and time.

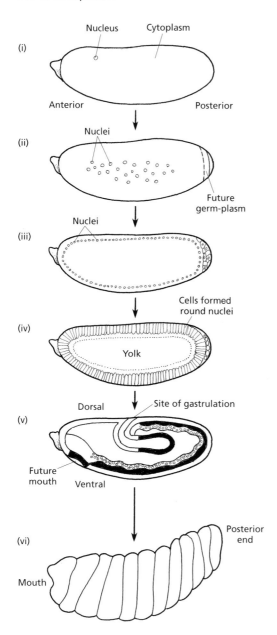

Figure 19.4. The early development of *Drosophila*: (i) the egg; (ii) syncitium formed; (iii) nuclei at the periphery; (iv) cytoplasm rounds up around nuclei, making cells; (v) gastrulation; (vi) external view of segmented larva.

Not only must cells divide, move and differentiate, but the products must be arranged in patterns. The first step in pattern formation is the determination of polarity, establishing **the principal axis** of the embryo. In *Drosophila* the head end is defined in the unfertilised egg, under instructions from maternal messenger RNA.

19.5.1 The dorsoventral axis

The dorsoventral axis is laid down at right angles to the first axis, either in response to the point of sperm entry at fertilisation or during unequal cleavage. Yolk is heavy, so that owing to gravity the yolk-containing 'vegetal' region becomes the lower part of the egg and the nucleus remains in the upper 'animal' part. In sea urchins the maternal RNA as well as gravity determines the localisation of yolk in the egg. The mouth forms in the animal part of the developing egg and the vegetal pole becomes posterior. In gastropods, neither axis is established until early cleavage, when the large D blastomere marks the future posterodorsal region (see Figure 19.1). In nematodes the first cleavage divides the egg unequally, into a large blastomere, AB, which becomes anterior and a smaller one, P, which becomes posterior (see Figure 19.3). The next cleavage establishes the dorsoventral axis. As has been emphasised in Chapter 8, the unchanging cell lineage of nematodes is very important, but experiments show that cell interactions are also involved: if AB is rotated, the dorsal axis then forms upside down, not only in AB but also in P.

19.5.2 Body symmetry

Most animals show bilateral symmetry, established by the first two axes and definition of right and left sides. The single axis of the radially symmetrical cnidarians and ctenophores appears during cleavage.

19.5.3 Positional information

Two axes at right angles to each other can set up a coordinate system giving each cell a positional identity. In *Drosophila* both axes are laid down in the unfertilised egg, and after fertilisation the RNA localised at the anterior end is translated into a protein that diffuses along the anteroposterior axis, forming a concentration gradient. Such a chemical involved in pattern formation and acting differently at different concentrations is called a 'morphogen'. Typically it passes with decreasing concentration from a source at one end to a sink at the other end, and cells along the line are in this way provided with positional information and develop accordingly.

19.6 What happens in later development?

Later development continues the processes of cell division, growth, cell movement and commitment followed by differentiation. There are series of **inductions**, where signals produced by one cell population change the developmental fate of another group of cells. The morphogens involved in axial patterning provide an early example. H. Spemann and O. Mangold in 1924 discovered what they called the 'organiser', which, during amphibian gastrulation, induces neural development in ectodermal cells with which it comes into contact. In invertebrates also, induction underlies the determination of the cell layers; for example, in sea urchins the smallest blastomeres at the vegetal pole determine cell fate as gastrulation occurs. In most animals, cell-to-cell interaction is even more important than cell lineage in determining the course of development. Progressive induction defines the fate of cells with increasing precision and produces an increasingly elaborate embryo.

The life history then takes one of two patterns: either the embryo hatches early as an immature form, a larva, or it develops directly into an adult. The forms and significance of larvae (see Box 17.1) have already been introduced, especially in crustaceans and echinoderms (Chapters 13 and 17).

19.7 What can studies of regeneration tell us about development?

Development need not start with an egg. Asexual reproduction is frequently found, and most animals can regenerate lost tissue on a small scale. Many can regenerate the whole body from a small piece (some sponges and cnidarian polyps can do this even after the cells have been separated by sieving). The whole animal can be regenerated from as little as one-sixteenth of the body in some planarians, or from blood cells alone in ascidians. Studies of regeneration have illuminated our understanding of development. An example from planarians is given shortly, and two examples, both from cnidarian polyps, are as follows:

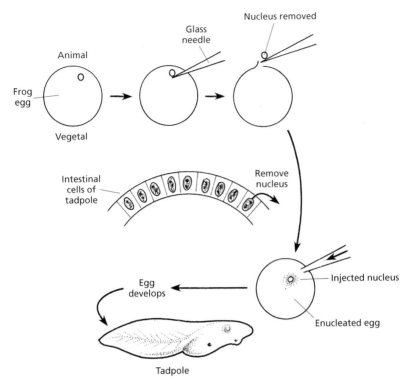

Figure 19.5. Nuclear transplantation in the frog.

- *Hydra*: morphogens activating and inhibiting head and foot formation have been identified as peptides stored in neurosecretory granules in nerve cells. If the 'head' is removed, these morphogens establish a new high point for the gradient along the head to foot axis, and determine where regeneration can occur. The original concept of positional information was derived largely from studies of the gradients in regenerating *Hydra* (see Figure 19.6*a,b,c*).
- *Tubularia*: a colonial hydroid with large hydranths bearing two whorls of tentacles. When the polyp develops from the planula larva, these tentacle develop from base to tip, starting as knobs that elongate progressively. When, however, the hydranth is cut off and the cut end regenerates, tentacles are folded off laterally from the stem (see Figure 19.6*d*). This is a striking example of development of the same structure by two different routes, and at once points to the flexibility of the gene control of development.

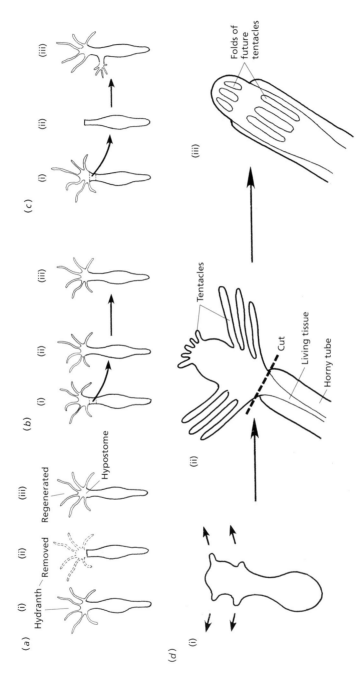

Figure 19.6. Regeneration in hydroids: (a) Hydra, removal and regeneration of the hydranth; (b) hypostome graft into intact Hydra, no regeneration; (c) hypostome graft into beheaded Hydra, a new axis is formed; (d) Tubularia, (i) tentacle growth in embryo, (ii) adult hydranth, (iii) regeneration.

19.8 How do genes control development?

This is a vast question! But modern work is beginning to give us answers. Development from the egg always requires thousands of genes and a complex regulatory programme that organises the expression of genes in time and space. Our information comes primarily from mutants. What goes wrong when a gene is abnormal or absent may help us to identify its normal function (which is why a gene may be called 'eyeless' when its function is to make eyes). The fruitfly *Drosophila* is the main source of information because no animal's mutants are better known. The surprise is that many animals turn out to control their development with genes closely related to those known from *Drosophila*. That many genes have changed so little during evolution is one of the most remarkable discoveries of recent years; change occurs primarily in the regulatory interactions that control genes. Genes regulating early development are likely to have especially widespread effects, and these genes are often found to have similar roles in many phyla. There are other regulatory mechanisms; for example, in the nematode *Ascaris* certain genes become lost in early divisions of all cells except the germ cells. In general, however, development is controlled by the mechanism regulating the expression of genes, not by altering the gene content of cells. Morphological diversity arises from changes in this regulating process.

Genes that choose between alternative pathways in early development are called **selector genes**. They produce an orderly cascade of genetic instructions for the developing animal, and a small change in a selector gene may produce a large change in the animal. One class of selector genes is defined as 'homeotic', meaning 'becoming like': the term was first applied to mutants of *Drosophila* in which one part of the animal becomes like another: for example, 'antennapedia', where a foot is formed in place of an antenna; and 'bithorax', where the third as well as the second thoracic segments bear a pair of wings (as in the four-winged ancestors of the true flies such as *Drosophila*).

19.9 What are *Hox* genes?

Hox genes determine the relative position of structures along the major body axis. *Hox* genes were discovered by studying homeotic mutations in *Drosophila*, where they determine the number of segments and the number and sequence of appendages and initiate a hierarchy of position-regulating genes that organise further development. They have been remarkably conserved in evolution: the same genes occur in vertebrates and all the invertebrates that have been examined.

Hox genes belong to a large family of regulatory genes called **homeobox genes**. The homeobox is a sequence of 180 nucleotide base pairs coding for a conserved sequence of 60 amino acid residues called the **homeodomain**. Interaction of homeodomains with target DNA sequences determines whether or not those target genes are expressed.

Hox genes share three characteristics. Firstly, they have closely related homeoboxes and share some other aspects of protein sequences. Secondly, they are organised into clusters, which probably arose very early in the evolution of animals from the lateral duplication of an ancestral *Hox* gene. Thirdly, *Hox* genes are always expressed in the same order along the anteroposterior axis and, remarkably, are also arranged along a chromosome in the order corresponding to their sequence of expression.

19.10 How do *Hox* genes act in different animals?

The first impression is one of uniformity. Members of all animal phyla have genes that specify cell types and relative positions within the body. In animals, these genes often contain the homeobox (whereas plants and fungi use other classes of regulating genes for most of these decisions). The set of *Hox* genes, and other regulating genes specific to animals, has been termed the 'zootype', and may be a defining characteristic of animals as a whole. The zootype is most clearly displayed after gastrulation but before cell differentiation. This, rather than any earlier developmental stage, is when all the members of a phylum show a maximum similarity to each other. Called 'the phylotypic stage', it embodies the first appearance of the body plan characterising a particular phylum.

Hox genes and the proteins that they encode have been remarkably conserved during evolution. Developmental changes arise, however, when *Hox* genes are duplicated or lost, or when their clusters change in size or number. Above all, there have been changes in their relationships with other genes. Once again it is the regulation of gene expression, rather than the genes themselves, that has changed during evolution. *Hox* genes are at the apex of a hierarchy of regulator genes. For example, in *Drosophila* there are eight *Hox* genes among two complexes, known as the antennapedia and bithorax complexes (see above for the action of the mutants giving them their names). These *Hox* genes are expressed as regulatory proteins that determine the plan of the body segments and the sites and types of appendages. They have a wide range of target genes 'downstream'; for example, one *Hox* gene (ultrabithorax, *Ubx*) regulates at least 80 other genes. In different species of insects, the same *Hox* gene may have different target genes, and in different groups of arthropods it may be expressed in different segments with quite different target genes controlling the pattern of appendages. While arthropods are the invertebrates in which *Hox* genes have been most thoroughly studied, information from other phyla helps us to understand the action of homeobox-containing genes during development. For example:

— **Platyhelminths:** *Hox* genes have been studied in regenerating planarians. Cell division throughout the body is an initial response to wounding, and at the same time *Hox* genes are expressed in sequence, so that the new cells (which at first cannot distinguish head from tail) are able to acquire positional information. At the end of regeneration, *Hox* genes are 'switched off', again in sequence.

— **Nematodes:** although these worms have neither segments nor appendages, the patterning of the anteroposterior axis is similar to that of *Drosophila*. One nematode *Hox* gene called *mab5* is normally expressed in the posterior region. If these posterior blastomeres are moved experimentally, *mab5* is then expressed by these cells regardless of their new position: as so often in nematodes, cell behaviour is determined primarily by cell lineage.

— **Echinoderms:** the control of early development in sea urchins was first described in terms of interacting gradients from the animal and vegetal poles. These gradients have subsequently been reinterpreted as successive activation of particular sets of genes. Bilateral symmetry is, however, only temporary: the axis laid down in the larva is superseded by the five-rayed symmetry of the adult. Most *Hox* genes are not used during development of the larva, but are expressed in the adult in ways specific to the phylum, such as development of the tube feet and the skeleton. This is an example of highly conserved genes being used in new ways; details of their use differ between classes of echinoderms.

— **Chordates:** comparisons between the amphioxus and vertebrates show most strikingly that duplication of *Hox* gene clusters can affect development.

Chapter (20) shows that study of *Hox* genes not only explains some developmental processes but also illuminates our understanding of evolutionary relationships.

19.11 Conclusion

Development is a process with emergent properties. An egg has potential, but until it has divided its parts cannot move around. After cleavage, gastrulation shows that sheets of cells have properties that separate cells do not have. At the present time our knowledge about the gene control of development is growing very fast indeed, as the orderly sequence of gene action becomes revealed. It is exciting to discover how development is initiated and directed by gene action, but the process cannot entirely be explained in these terms. The behaviour of cells, like the behaviour of whole organisms, cannot entirely be predicted from the genes that they contain. Interactions between the component parts, and between the developing animal and its environment, will affect the end-product.

20

Invertebrate evolutionary history

The concluding chapter attempts to draw together this introduction to invertebrates by discussing their evolutionary relationships (phylogeny).

20.1 How can we trace the course of evolution?

1. **Morphological evidence** has long been used to establish animal relationships, even before the evolutionary basis for these relationships was recognised. Modern methods of structural analysis, for example electron microscopy, have greatly improved our understanding of morphology. Yet owing to the prevalence of convergent evolution, morphological comparisons alone cannot give us a certain guide to the course of evolution, as was explained in Chapter 2, 'The pattern of evolution'. Chapters 3–18 illustrate this fact and Chapter 19, 'Development', establishes that the morphological interpretation of early development is subject to the same uncertainty.

2. **Fossils** provide morphological evidence about animals that lived long ago. Recent finds, in particular finds of very early animals, have greatly added to what we know. The first part of this chapter briefly introduces this vast topic.

3. **Molecular evidence**: since the late 1980s molecular biology has provided a new tool for assessing the pattern of evolution, using molecules that have changed very slowly over evolutionary time. Molecular evidence is not inherently preferable, but it is valuable because it can be compared with morphological evidence from both living animals and fossils. The

last decade of the twentieth century saw an enormous output of papers about molecular phylogeny; this chapter indicates the main results and the preliminary conclusions.

20.2 What are fossils?

20.2.1 Fossils

Fossils are the remnants of once-living animals, preserved in rocks or (less often) in sediments, amber, ice, etc. The term is normally reserved for remains dating back to before the last Ice Age. Hard parts of animals may be preserved with little change in appearance: among invertebrates these include the shells of molluscs and brachiopods, echinoderm skeletons and spicules, jaws, etc. from many phyla. Arthropod exoskeletons may be fossilised whole or preserved as thin films of carbon on rocks. Fossilised soft parts have usually been turned to stone by replacement of organic material by minerals from solutions underground.

20.2.2 Living fossils

This is a potentially confusing description of animals which have changed remarkably little over long periods of time. Examples include the brachiopod *Lingula* and the horseshoe crab *Limulus*.

20.2.3 Trace fossils

Trace fossils are imprints on the environment of the activity of animals long ago, such as tracks, trails, burrows, coprolites (i.e. fossilised faeces) and impressions on soft substrata that may be made even by animals such as jellyfish.

20.3 What can fossils tell us about evolution?

The fossil record can tell us about the structure and way of life of past animals and the sequence in which they appeared: the facts about the time dimension may enable us to root evolutionary trees. Obviously the fossil record is very

MYA: Millions of years ago, to the beginning of each period.

CENOZOIC: 65 Tertiary   Pangea drifted apart, birds and mammals spread.

MESOZOIC: 140 Cretaceous   Ended at K/T border with extinctions of dinosaurs, ammonites and much else.

240 Jurassic   Earliest birds and mammals.

250 Triassic   Teleost fish, dinosaurs appear.

PALAEOZOIC: 330 Permian   Continents join as Pangea, including Gondwanaland. Climate deteriorated, loss of diversity, extinctions of trilobites and many brachiopods and ammonites. Ended in an Ice age.

360 Carboniferous   Warm seas, food plentiful, great diversity. Amphibia, reptiles, giant dragonflies. Age of forests, trees with seeds.

410 Devonian   Old Red Sandstone, horsetails. Graptolites gone. 'Age of Fish', all kinds, and first tetrapod vertebrates.

430 Silurian   Making of mountains. Land colonised by plants and some arthropods.

500 Ordovician   Continental drift; more food, PO_4, O_2? Trilobites, graptolites, nautiloids, coral reefs. Many Cambrian oddities replaced by familiar shells. Equator through Canada, Greenland, Siberia and Queensland, Australia (mapped by trilobites). Ended in an Ice age.

500+ Cambrian   Small, shelly fauna appear, phosphatic by replacement. 544, Cambrian explosion, all main phyla appear. Animals diversified and increased in size beyond millimetres. Rise of predators and animal interactions. (Evidence, 520 Burgess Shale, 530 Chengjiang, 540 North Greenland).

Earlier, Precambrian   Many groups of small animals. (Evidence, 560 Ediacaran, 680 South China).

First animals. When?

3500   Oldest microscopic fossils. Tiny puffs of oxygen. Continents.
3800   Earliest preserved rocks, Greenland.
4600   Earth spun off sun, plus accretions from meteorites: carbon.

Figure 20.1.   The geological succession.

patchy and very incomplete: animals lacking hard skeletons are less likely to leave fossil evidence. Even where a group has as many fossils as the arthropods, some animals once very common (such as the trilobites) occur in large numbers while rarer forms may not be represented at all. Nevertheless, not only can the fossil record tell us about the morphology of past organisms, it can also with careful interpretation reveal facts about animal mechanics and about ecological interactions and the nature of past ecosystems.

The succession of life forms that can be traced is sufficiently reliable for geologists to date rocks by the fossils which they contain (Figure 20.1). As the sequence in which the main groups of animals occurred is determined, new finds continually push the earliest appearance of each group back in time. Direct ancestry, however, is not revealed. While every fossil must have a nearest living relative, we can only very rarely identify it, and, in any case, the chance of finding a direct ancestor is vanishingly small. Among fossils as among living animals, a supposed 'missing link' between phyla can seldom be authenticated, although it may be disproved (for example, we know that the onychophoran *Peripatus* is not a link between annelids and arthropods, see below). Yet where fossils are plentiful a group of intermediate and possibly transitional forms can sometimes be identified – information that molecules can never provide.

Barnacles on the seashore give us a nice example of such information. From the earliest fossils to modern forms there is a trend of reduction of the number of lateral plates round the body. The rich fossil record enables us to trace this reduction in seven of the eight lineages of barnacles in this family, a remarkable example of parallel evolution. Why should it have occurred? Fewer plates means fewer junctions between plates, and observations of living animals show that common predatory gastropods attack barnacles at these junctions. The plate reduction coincided with the rise of this family of predaceous gastropods in the Cretaceous. Further, the one lineage with no plate reduction is *Chelonobia*, barnacles living on turtles and free from gastropod predation !

20.4 Can we date the earliest appearance of animals?

The difficulty is that all the main phyla seem to have appeared over a relatively short period of time in the Cambrian. Modern methods (based on the rate of breakdown of uranium to lead) date the 'Cambrian Explosion' as beginning 554 million years ago (MYA) and lasting for 5–10 million years. Our knowledge of early animals has been derived mainly from five very rich fossil sources; in order of increasingly distant past time, they are:

– **The Burgess Shale in Canada** (520 MYA): the steeply sloping hillside leading down to a toxic, oxygen-depleted sea bed probably caused animals to be buried rapidly and protected at the bottom from scavengers and from bacterial decomposition. This fossil assemblage, first discovered early in the twentieth century but only recently fully analysed, has more than any other formed our picture of the Cambrian fauna.
– **Chengjian, Southwest China** (530 MYA) has recently filled in that picture,

with many of the same animals in different proportions. In particular, the well-preserved arthropods have given us a firm idea of their early diversity.

– **Sirius Passet, North Greenland** (540 MYA) more recently still has, among other riches, provided fossils of what may be stem arthropods. This suggests that arthropods originated in the Precambrian.

– **The Ediacaran Range in South Australia** (560 MYA) has been known since the mid twentieth century. These Precambrian (now Lower Cambrian, the border moves back) fossils were originally thought to have evolved quite separately from the Metazoa. They were called 'vendobionts', after the Precambrian Vendian period. It is now clear that they are not separate but consist of many early metazoan forms, not only cnidarians but a number of other phyla such as molluscs and various worms. Probably they were all soft bodied and depended for their food on symbiotic microorganisms (see Box 16.1).

– **The Yangtse Gorge, South China** (580 MYA and earlier) has phosphorite deposits that very recently have provided us with most striking early fossils. In addition to undoubted sponges, revealed by cellular imprints as well as spicules, there are early stages of embryonic cleavage, two and four cell stages and later ones with many blastomeres.

Why did the Cambrian explosion occur? Explanations dwell on increasing oxygen supplies, more possibility of making collagen and hard structures and the associated increase in interactions between species, particularly the rise of predators.

The 'Cambrian Explosion' is a correct description of the fossil record as we have it, and also of the difficulty of resolving these basic relationships by molecular evidence. Yet the different phyla (or even groups of phyla) must have separated considerably earlier. Dating origins is very controversial at present: molecular methods suggest much earlier dates for the establishment of animals, and that the earliest divergence between diploblastic and triploblastic animals may have occurred 1000 MYA or more. Is this because molecular change may not accurately measure the passage of time when evolution is rapid, or is it because earlier fossils have not been found? Evolutionary lines could have separated long before morphological differences became detectable in fossils; perhaps the earliest animals were too small to be recognised among the single-celled fossils known from the very early rocks. We await more evidence.

In conclusion: fossils must be included in phylogenetic reconstruction, and can influence the placing of present-day groups. We do risk overemphasising fossilised characters, but as with any method, we have to work with what we have. We do risk interpreting fossils according to our preconceptions but again, we can do that in any form of biological enquiry. At least we can now compare the evidence from fossils and molecules to assess the relationships of present-day animals.

20.5 What do molecules tell us about relationships between phyla?

Genes are remarkably conservative – that has been one of the main surprises of the last two decades, and the reason why recent molecular work has concentrated on genes. The early concept of genes as either 'structural' or 'regulatory' has been modified by the discovery of a few highly organised gene families controlling developmental processes. Genes patterning development act on networks of other genes described as 'downstream' from them; the controlling genes have changed remarkably little during the course of evolution – perhaps any change would be selectively disadvantageous – and what has changed is the downstream network.

A striking example of gene conservation is the *pax6* gene, which is expressed at the earliest stage of eye development. The protein specified by *pax6* initiates eye production in animals as different as insects, squids and ourselves, and also in animals with minimally developed eyes such as platyhelminths and nemertines. Insertions of *Drosophila pax6* (gene or protein) into limb rudiments can induce eyes in the wrong place, and insertions of squid or mouse *pax6* can have the same effect. When the squid *pax6* gene is grafted into the olfactory organ, brain or arms of a squid, it is expressed, but with consistent downstream differences. Arthropod compound eyes have long been considered to have evolved quite independently of the vertebrate simple eye: this 'independence' now needs qualification. The remarkable resemblance between the eyes of cephalopods and vertebrates is quoted as a classic case of convergence: this remains correct, only the convergence is based not on the gene initiating eye development but on genes downstream from the initial stimulation.

Much has been revealed about evolutionary relationships from studies of other highly conserved genes, in particular the genes coding for ribosomal RNA (see Chapter 2) and the *Hox* genes that pattern development along an axis (see Chapter 19). Research on these genes and others is exceedingly active, so that any publication of the results will become out of date very soon.

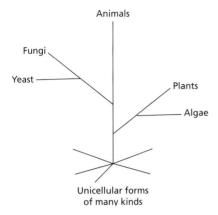

Figure 20.2. The relationships between kingdoms of living organisms.

Nevertheless, here is the picture of the course of animal evolution that is widely accepted at the outset of the twenty-first century.

20.5.1 Animals

Animals are more closely related to fungi than either is to plants. Figure 20.2 represents one possible sequence, or animals and fungi may have separated much earlier.

20.5.2 Metazoa are monophyletic

All the multicellular animals we know today had a common ancestor. This is suggested by the uniform molecular constitution of the extracellular matrix, the ubiquity of collagen and many other biochemical features, as well as by 18 S rDNA. *Hox* genes provide even more compelling evidence that sponges and cnidarians did not evolve quite separately from other animals.

– **Sponges**, according to traditional morphology, are the simplest multicellular animals, and they are the earliest branch to leave the tree constructed from ribosomal gene data. Since adult sponges have no axis, no *Hox* genes can be expected, yet homeobox-containing genes are found even here, as indeed they are in yeast: their function is not clear.

- **Ctenophora**: this phylum also branches off the main line separately and early.
- **Cnidaria** branch off after sponges, perhaps after ctenophores. Here *Hox* (or '*Cnox*') genes, similar in part to those of triploblasts, define the main axis. They have effects on regenerating *Hydra* parallel to those of certain *Drosophila Hox* genes (a finding that justifies calling the oral end of *Hydra* a 'head').

Clearly a small region of a gene sequence has been conserved over a remarkably long period of time – perhaps any mutation would be too disruptive of development. It is this uniformity that has led to the definition of an animal as an organism that displays a particular spatial pattern of gene expression, the 'zootype'.

20.5.3 Triploblasts

Triploblasts triggered the Cambrian explosion. The timing of these events, as has been explained, is very controversial, with molecular evidence suggesting origins considerably earlier than fossil evidence can verify. Rapid radiation is a particular problem for ribosomal RNA genes, which cannot resolve time differences of less than 40 million years. Too much weight should not be put on evidence from a single gene. Yet a study of 18 other genes placed the division of the triploblasts as early as 670 MYA.

20.5.4 Protostomes and Deuterostomes

Molecular evidence supports this traditional deepest division of the triploblastic animals, even though there are anomalies and apparent intermediates in the early developmental features defining the division (see Chapter 16).

Protostomes
These include most triploblasts.

Platyhelminthes and Nemertea The theme of Chapter 6 is that the traditional view of platyhelminths as basic triploblasts obscures the specialised nature of these modern animals. Whether or not the phylum is monophyletic (with or without the acoelans), platyhelminths (and also nematodes) have been hard to place on account of the very rapid evolution of ribosomal genes ('long branches' on the constructed trees) in many examples. Evidence from 18SrRNA genes rests on subjective selection of species with slow evolution ('short branches'). Such evidence does firmly place the platyhelminths among the protostome coelomates rather than at the base of the triploblasts; more recently, evidence from *Hox* genes has confirmed this placing, which applies to nemertines also. That they are close to coelomates of course tells us nothing about the direction of evolution: it certainly does not imply that platyhelminths are coelomates that have lost their body cavity (there is a great deal of evidence to the contrary) nor that the nemertine rhynchocoel and mesodermal blood system are remnants of an ancestral coelom. Support for the claim that nemertines are reduced coelomates would require much evidence that at present does not exist.

Annelida and Mollusca The traditional placing of these two phyla together, on account of similarity in early embryology and larval form, is supported by molecular and also fossil evidence, for example the Middle Cambrian.

Wiwaxia This fossil has outgrowths half way between molluscan scales and annelid chaetae. Annelids are unlikely to be monophyletic: 'polychaetes' are probably even more diverse than had been realised before. Any combination of annelids with arthropods as 'Articulata' (both being segmented) has been abandoned by most morphologists and is not supported at all by molecules.

Phoronida, Bryozoa and Brachiopoda Radial cleavage and regulative development suggest deuterostome affinities for these animals, yet subsequent features of their early development are either ambiguous or clearly protostome (see Chapter 16). The suggestion that radial cleavage may be primitive among metazoans, with spiral cleavage a secondary specialisation, further devalues a diagnosis based on radial cleavage alone.

Combination of fossil and molecular evidence can resolve this uncertainty about phylogeny. Brachiopods are very common in the fossil record, and resemblances to various early protostomes abound: for example, unlike deuterostomes brachiopods have chitin, and their mantle-edge bristles are identical in ultrastructure to polychaete chaetae. There are Lower Cambrian fossils called halkieriids that are in several ways intermediate between annelids and molluscs on the one hand and brachiopods on the other. One intriguing slug-like halkieriid from North Greenland has scales like the mollusc *Aplysia*, with a larger shell at either end of the animal; it is tempting to picture this form contracting to make a brachiopod. While convergence cannot be ruled out, fossils suggest that annelids, molluscs and brachiopods are closely related.

Molecular investigation of modern animals strongly supports protostome placing for brachiopods, bryozoans and phoronids, using the 18SrRNA genes and (more recently) *Hox* genes: a posterior *Hox* gene group from the brachiopod *Lingula* is unlike any deuterostome genes and not close to arthropod ones, but is very similar to posterior *Hox* genes in the annelid ragworm *Nereis* and the limpet (mollusc) *Patella*.

Nematodes Most surprisingly, nematodes are now placed near the arthropods, reviving an ancient grouping of animals with hard cuticles that need to moult, and use ecdysone in the process. This grouping does not only rest on choice of nematodes with slowly evolving ribosomal genes: it is strongly supported by the study of several *Hox* genes, which are known in great detail both for the nematode *Caenorhabditis elegans* and for the arthropod *Drosophila melanogaster*.

Protostomes accordingly are divided into two groups, the 'Lophotrochozoa' and the 'Ecdysozoa': Lophotrochozoa include annelids, molluscs and protostomes with lophophores (phoronids, bryozoans and brachiopods), and Ecdysozoa include all arthropods, onychophorans, tardigrades and nematodes. The platyhelminths and the nemertines are close to the Lophotrochozoa: opinions vary as to whether they should be included there.

Re-examination of 'Aschelminthes' This is now essential. A 'pseudocoelom' can be formed in various ways, and may simply be due to the absence of a mesodermal body cavity. Absence of a character cannot unite disparate groups, and the union of many small phyla as 'pseudocoelomates' has long been considered unsatisfactory, as was explained in Chapter 5. Molecular evidence provides some hope of positive information about their relationships.

The phyla introduced in Chapter 5 are at present placed as follows, mainly from 18SrDNA evidence:

– **Lophotrochozoa** include **Rotifera, Acanthocephala, Gastrotricha** and **Gnathostomulida** (some or all of these phyla are perhaps nearest to the platyhelminths). An interesting addition to these triploblasts is the misnamed phylum **Mesozoa**. The dicyemids (rhombozoans) can be related to Lophotrochozoa both by ribosomal genes and by one *Hox* gene. Clearly these very simple endoparasites are not basal metazoans but secondarily specialised. **Myxozoa**, formerly misdiagnosed as protistans, are similarly revealed as greatly simplified triploblasts, containing *Hox* genes. As in sponges, their function in forms with no body axis is obscure.

Of the coelomate phyla, **Sipuncula** are close to molluscs and both

Pogonophora and **Echiura** probably arose from within the annelids, which are confirmed as not being monophyletic. **Chaetognatha**, a small phylum notoriously hard to place, may prove to be closer to the Ecdysozoa, but further evidence is needed.

— **Ecdysozoa** appear to include all the moulting 'aschelminthes', i.e. **Nematomorpha** as well as nematodes, **Kinorhyncha**, and probably **Loricifera**. **Priapula** are identified as belonging to this group not only by 18SrRNA but also by some posterior *Hox* genes, which they share with nematodes and arthropods but not with brachiopods, annelids or molluscs.

These results are summarised in Figure 20.3.

— **Deuterostomes**: the only remaining deuterostome phyla are **Echinodermata**, **Chordata**, and **Hemichordata** (if they are a separate phylum). The relationship between them, and in particular the nature of hemichordates, is as controversial as ever. These relationships have already been discussed in Chapter 18 because the problems are integral to any consideration of the invertebrate chordates; molecular investigations are being made but these problems are as yet unsolved. What has been illuminated is the origin and evolution of vertebrate chordates. The old (1822) suggestion that vertebrates were 'worms turned upside down' with the nerve cord becoming dorsal and the main blood vessel ventral has been rescued from ridicule by study of the genes specifying the dorsoventral axis; also the progressive increase of complexity from the body form represented by the amphioxus to jawless vertebrates (and then to vertebrates with jaws) has been associated with duplications of *Hox* gene clusters from one to two to four. Vertebrates, however, are outside the scope of this book.

What, then, were the earliest triploblasts like? Suddenly we have lost the 'flatworm' precursor. Could they even have been coelomate segmented animals, with the coelom and segmentation lost in many phyla? This does seem improbable, but is currently the subject of much research and debate.

20.6 What do molecules tell us about relationships within phyla?

Chapters 3 to 18 include discussion about the relationships between different classes within each phylum, where this can be ascertained, with brief reference to molecular as well as morphological evidence. Two examples are

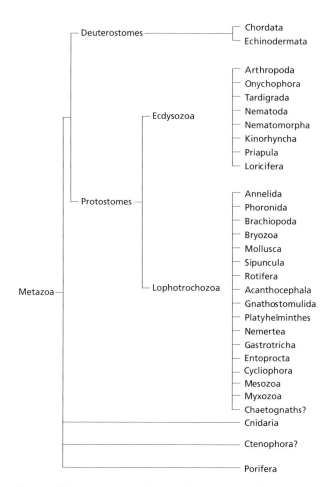

Figure 20.3. A recent scheme of metazoan evolution.

considered more fully below: echinoderms because the fossil record is particularly good for many classes, and arthropods because the *Hox* genes of many examples are becoming particularly well known. In both these examples molecular evidence can be compared in detail with morphological evidence, both from fossils and from living animals.

20.6.1 Echinoderms

The unique skeleton can be found in fossils and shows characters known from present-day forms. Fossil evidence tells us that echinoderms were present in

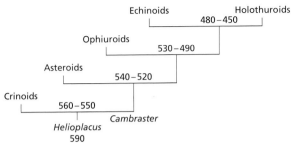

Figure 20.4. The relationships between echinoderm classes, based on fossil evidence. The numbers represent millions of years ago (MYA) and record the earliest possible times of divergence (higher figures) and the times when divergence has certainly occurred (lower figures). The Cambrian began at 600 and ended at 500 MYA.

the early Cambrian (590 MYA) and suggests possible relationships between the five main classes. Figure 20.4 traces their divergence up to their complete separation, 450 MYA in the Ordovician. The estimated time of each divergence is shown as two figures: the upper limit defines when separation is certain and the lower limit gives the earliest possible time of separation.

The history of sessile echinoderms, including the crinoids, is well known. Relationships within free-living Classes are known best for the echinoids, where fossils are particularly abundant (10 000 species have been described) and where the same characters can be used for classification as in the 900 known present-day species. The rates of morphological change can be estimated, and correlate well with time elapsed, over 250 million years. Ophiuroid trees are well rooted and supported by fossil evidence; asteroid trees much less so, and some authorities place asteroids closer to ophiuroids than Figure 20.4 suggests. Holothuroids, with poorly developed skeletons, have left few fossils and their relationships remain controversial.

Molecular evidence, largely derived from genes specifying ribosomal RNA, has not always given results agreeing with the fossil evidence. However, relationships within the echinoids have now been clarified by a particularly thorough comparison of these two sources of information, comparing 163 characters of fossil morphology with analysis of two different ribosomal genes (those for 15 S and 28 S rDNA). In addition, familiar *Hox* genes have been found, expressed in ways specific to echinoderms. Three of these genes appear to be responsible for several of the unique features of the phylum. The expression of echinoderm genes downstream from *Hox* is different in the different Classes: the genes may even be expressed at different regulatory levels. For example, 'distal-less', which, in *Drosophila*, controls distal limb development, is expressed in the tube feet of asteroids, echinoids and holothuroids but not in those of ophiuroids, where a different gene (also known from *Drosophila*) controls development of tube feet. In asteroid larvae, distal-less is expressed in ectodermal cells and in particular in the brachiolar arms, used for attachment. In echinoids, distal-less is expressed from the first appearance of the adult rudiment and specifies most of the structures on the oral side.

20.6.2 Arthropod monophyly and origins

Morphological study of living groups can provide arguments either for a single origin of arthropods (monophyly) or separate origins for crustaceans, chelicerates, insects and others (polyphyly), but both fossil evidence and molecular evidence strongly support monophyly. For example, the biramous limbs of crustaceans were supposed to set them apart from arthropods with unbranched limbs, which were united as 'Uniramia' (including onychophorans, myriapods and insects). However, study of Permian fossils now reveals that the limbs of the earliest arthropods had many branches, the outer ones often subsequently lost. Modified outer branches of the primitive limb can be traced in most groups of arthropods, supporting the body wall (in trilobites, chelicerates and others), forming gills (in aquatic chelicerates and crustaceans) and, most strikingly, forming the wings of insects. The patterns of gene expression controlling these structures gives detailed support to the fossil evidence of their origins.

In general, molecular evidence from many sources strongly supports a single origin for arthropods, but in which group was this origin, and what is the position of onychophorans (represented today by terrestrial *Peripatus*, see Figure 12.6) now that the group 'Uniramia' has been demolished? Myriapods and insects are true arthropods: onychophorans are not. Recent fossil finds in Greenland and South China suggest that the stem groups of both arthropods and onychophorans may be traced back to Cambrian marine animals called 'lobopods', whose general body form is not unlike that of *Peripatus*, except for the presence of gills. Present-day tardigrades may well be related to this lineage. Figure 20.5 indicates a possible phylogeny.

20.6.3 The main groups of arthropods

The placing of the main classes of arthropods in Figure 20.5 is very preliminary: for one thing, while insects (Hexapoda) and Chelicerata are almost

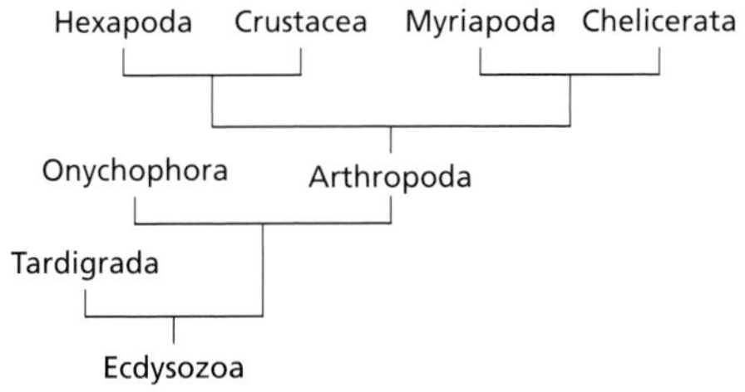

Figure 20.5. A possible phylogeny for the arthropod Classes.

certainly monophyletic, Myriapoda may be polyphyletic. Crustacea as at present defined also may not be monophyletic; there are suggestions that insects arose from among them.

Insects and crustaceans

Evidence from two distinct ribosomal genes supports the hypothesis that insects and crustaceans have a common root within the arthropods. Recent work on the control of segment development by *Hox* genes in the insect *Drosophila* and the crustacean *Artemia* illuminates this relationship, and is briefly introduced here to show how such information may underpin our phylogenies in the future. In *Drosophila*, the different *Hox* genes expressed at different points along the anteroposterior axis of the embryo are well known. They regulate the activity of 'downstream' genes that determine the differences between segments. In *Artemia* the head segments correspond to those of the insect, except that downstream genes allow development of a second pair of antennae on the appropriate segment. Behind the head, all the segments of *Artemia* are alike, although the *Hox* genes present are the same as in the insect: the difference is due to different patterns of expression of two of the genes mentioned in Chapter 19, ultrabithorax (*ubx*) and antennapedia (*antp*). For example, in immature *Artemia* (and in many other shrimps) these genes are fully expressed and no maxillipedes develop, but at maturity expression of these genes is withdrawn and maxillipedes appear. Study of *ubx* has revealed its role even in causing morphological differences (the patterns of hairs on the legs) between species of *Drosophila*.

A further example of downstream regulation is that 'distal-less' initiates development both of crustacean biramous limbs and of insect uniramous limbs. In the mandible, unlike other limbs, distal-less is not expressed in insects or crustaceans (though it is expressed in the jointed mandibles of myriapods). This accords fully with the morphological reassessment of the mandible and with fossil evidence that the mandible in both insects and crustaceans evolved from the whole limb by truncation of the axis.

Chelicerates

The division of the body into prosoma and opisthosoma develops similarly in all living chelicerates studied and is a unique feature of the phylum. The total absence of antennae is not due to absence of the first limb-bearing segment: expression of *Hox* genes both in spiders and in mites shows that, on the contrary, the prosoma as a whole corresponds to the head of insects and crustaceans, with chelicerae on the first appendage-bearing segment and pedipalps on the second. Four pairs of walking legs follow: the appendages used as jaws by many insects and crustaceans are used as legs by chelicerates.

Myriapods

That insects were derived from arthropods with many leg-bearing segments is well established, but these ancestors did not have to be close to modern myriapods, and all the molecular evidence suggests that they were not. Centipedes have the same set of *Hox* genes as insects and indeed all groups of arthropods; that these genes have not been used to make specialised segments does not relate them to insects in particular. Indeed, on other evidence centipedes appear to be closer to chelicerates than to insects and crustaceans.

The emerging picture is that arthropods are indeed a monophyletic phylum sharing common *Hox* genes and therefore having a common origin of segmentation, with early divergence of downstream genes that later led to the separation of the four main groups. The relationship between homeobox change and evolutionary change appears to be very close, with morphological diversity arising from different developmental use of highly conserved genes. *Hox* genes, discovered and first studied in relation to ontogenetic development, are now an exciting tool for investigating phylogeny.

20.7 Can we now define homology?

The basic problem in tracing relationships is the difficulty of distinguishing between homology and resemblance due to convergence (see Chapter 2). Use of molecular evidence does not solve this problem, if only because convergence may occur in molecules also, but it may help us to define homology – for that is where the problem begins. Homology is recognised where particular structures have been established and maintained by evolution, so that similarity in morphology reflects a common evolutionary origin. If, however, homology is used to construct a phylogeny and then homology is deduced from that phylogeny, the danger of circularity does need to be noticed! This is why a completely different set of data, molecules as well as morphology, is particularly valuable. Nor can a common developmental origin provide a firm definition, since conserved adult morphology must depend in part on stabilising interactions between cells after development is completed. Much evidence suggests that developmental similarity cannot guarantee homology, and that homologous (or even the same) structures can arise through very different developmental processes (for example, in regeneration, see Chapter 19). Homology cannot always be defined by common developmental pathways.

In more modern terms, can we evoke continuity of information in our definition of homology? Some gene sequences may be conserved over very long periods of animal evolution: we now have many examples such as the initiation of eye development by *pax6*. Flexibility and change reside in the associated networks of regulatory genes, which may differ as to whether they are expressed and in their activity or range. Morphological differences may depend only on regulatory genes. Arthropod segments and appendages provide another clear example of this. Throughout the phylum there is continuity both in structure and in genetic information; even between arthropod and vertebrate appendages there is genetic continuity, although there is no structural continuity (arthropod and vertebrate limbs are no more homologous than insect and bird wings). Perhaps it is not possible to provide a precise modern definition for a word of pre-Darwinian origin, even when it describes a concept central to evolutionary biology.

However, understanding molecular mechanisms does enable us to identify possible sites and processes of evolutionary change. Phylogenies based on changes (at rates that may vary) in a single gene must be uncertain, but further sources of information are being found at an increasing rate; for example, studies of the whole genome in mitochondria. In the two examples chosen above, study of echinoderms is showing us how fertile it may be to combine information from fossils and molecules, and study of regulatory genes in arthropods reveals new uses for molecular information.

20.8 Conclusion

It is exciting to have new tools for tracing invertebrate relationships. We can now imagine becoming able to trace the course of evolution with some certainty. A record of history, locked in very deep time, is no longer hopelessly inaccessible. However, the study of relationships should not distract us from the animals themselves. Natural selection has given us an enormous variety of animals. Their structure, function and behaviour can tell us more about evolution than we can learn from any phylogeny. The greatest challenge now is to conserve these animals in the threatened and threatening world of today. If we cut down rain forests, kill off coral reefs and poison other habitats on land, in fresh water and in the sea, animals will disappear. Humankind, like any other species, cannot live in the absence of many other forms of life. Conservation is the most urgent task for all biologists today.

Further reading

This book draws on a very large number of recent publications. This list is mainly selected under three categories: large-scale works of reference, books or reviews giving a general view of a topic and recent publications altering traditional ideas. Reference is confined to books and widely available journals such as *Nature*, London, or *Science*, New York. In addition, the Internet is increasingly a valuable resource.

General reference

Ruppert, E. E. & Barnes, R. D. (1994). *Invertebrate Zoology*, 6th edn. Saunders College Publishing, Philadelphia. A comprehensive recent textbook, with sections on general principles introducing each phylum.

There are many other large texts on invertebrates. Note: no book written before the mid 1990s can be up to date on phylogeny.

Pearce, V., Pearse, J., Buchsbaum, M. & Buchsbaum, R. (1987). *Living Invertebrates*, Blackwell, Palo Alto, CA. The enlarged successor to Buchsbaum, R. (1982). *Animals without Backbones*. Chicago University Press, Chicago.

Tudge, C. (2000). *The Variety of Life*. Oxford University Press, Oxford. An illustrated 'survey and celebration of all the creatures that have ever lived'.

Barnes, R. S. K. (ed.) (1998). *The Diversity of Living Organisms*. Blackwell, Oxford. Comprehensive and up to date.

Alexander, R. McNeil (1990). *Animals* (or its predecessor, *The Invertebrates*, 1979). Cambridge University Press, Cambridge. The mechanics of animal structure in relation to locomotion.

Harrison, F. W. and others (1991–1999). *Microscopic Anatomy of Invertebrates*. Wiley-Liss, New York. Specialist books (20 volumes) on the cellular structure of all invertebrates, up to date and invaluable. Not elementary reading except for very useful introductory summaries to some of the phyla and groups: worth seeking out.

General reading

New Scientist, weekly, is warmly recommended

Scientific American, monthly, has authoritative and beautifully illustrated articles, rather rarely about invertebrates

Gould, S. J. (1997). *Dinosaur in a Haystack*. Penguin, London. Also many other collections of essays.

Dawkins, R. (1976). *The Selfish Gene*. Oxford University Press, Oxford.

Dawkins, R. (1986). *The Blind Watchmaker*. Longmans, London.
Both Dawkins books are now in paperback (Penguin), as are many subsequent books by this author.

Conway Morris, S. (1998). *The Crucible of Creation*. Oxford University Press, Oxford. 'The Burgess Shale and the rise of animals.'

Fortey, R. A. (1998). *Life, An Unauthorised Biography*. Flamingo, London. 'A natural history of the first four thousand million years on earth.'

Wells, M. J. (1968). *Lower Animals*. Weidenfeld and Nicolson, London. Out of print and not up to date, but a pleasure if you can find it.

Wells, M. J. (1998). *Civilisation and the Limpet*. Perseus Books, Cambridge, MA. Read about animals, mostly in the sea.

Introduction to evolution and genetics (Chapters 1–2)

Jones, S. (1994). *The Language of the Genes*. Flamingo, London.

Jones, S. (1999). *Almost like a Whale*. Doubleday, London.

Wills, C. (1991). *The Wisdom of the Genes*. Oxford University Press, Oxford.

Ridley, Mark (1986). *Evolution and Classification*. Longman, London.

Ridley, Matt (1993). *The Red Queen*. Viking, London. (Reprinted 1994, Penguin, London.)

Panchen, A. L. (1992). *Classification, Evolution and the Nature of Biology*. Cambridge University Press, Cambridge.

Majerus, M. E . N. (1998). *Melanism; Evolution in Action*. Oxford University Press, Oxford.

Topics introduced in Chapters 3–18

Invertebrate groups

Simpson, T. L. (1983). *The Cell Biology of Sponges*. Springer-Verlag, New York.

Williams, R. B. (ed.) (1991). *Coelenterate Biology*. Kluwer, Dordrecht. A Symposium Volume published as a special issue of *Hydrobiologia*.

Dubinsky, W. (ed.) (1990). *Coral Reefs*. Elsevier, Amsterdam.

Taylor, J. T. (1996). *Origin and Evolutionary Radiation of the Mollusca.* Oxford University Press, Oxford.

Hanlon R. T. & Messenger J. (1998). *Cephalopod Behaviour.* Cambridge University Press, Cambridge.

Schram, F. R. (1986). *Crustacea.* Oxford University Press, New York.

Foelix, R. F. (1982). *Biology of Spiders.* Harvard University Press, Cambridge, MA.

Gee, H. (1996). *Before the Backbone.* Chapman & Hall, London. A survey of the chordates with 'Views on the origin of the Vertebrates'.

Particular topics in Chapters 3–18

Vacelet, J & Boury-Esnault, N. (1995). Carnivorous sponges. *Nature* **373**: 333–335.

Leys, S. P. & Mackie, G. O. (1997). Electrical recording from a glass sponge. *Nature* **387**: 29–30.

Plasterk, R. H. A. (1999). The year of the Worm. *BioEssays* **21**: 105–109.

Hodgkin J., Horvitz, H. R., Jasny, B. R. et al. (1998). *Caenorhabditis elegans*, sequence to biology'. *Science* **282**: 2011–2046. (See esp. pp. 2011–2018, The *C. elegans* sequencing consortium.)

Hadley, N. F. (1986). The Arthropod Cuticle. *Scientific American* **255**: 104–112.

Barlow, R. B. (1990). What the brain tells the eye (in the horseshoe crab). *Scientific American* **262**: 90–95.

Ellington, C. P., Van der Berg, C., Willmott, A. P. & Thomas, A. L. R. (1996). Leading edge vortices in insect flight. *Nature* **384**: 626–630.

Wootton, R. (1999). How flies fly. *Nature* **400**: 112–113.

Topics introduced in Boxes

Willmer, P., Stone, G. & Johnston, I. (2000). *Environmental Physiology of Animals.* Blackwell, Oxford.

Little, C. (1990). *The Terrestrial Invasion.* Cambridge University Press, Cambridge. The colonisation of land by invertebrates.

Barrington, E. J. W. (1979). *Invertebrate Structure and Function*, 2nd edn. Nelson, Sunbury-on-Thames. Not up to date but still informative.

Matthews, B. E. (1998). *An Introduction to Parasitology.* Cambridge University Press, Cambridge.

Development and phylogeny (Chapters 19, 20)

Wolpert, L., Beddington, R. S. P., Brockes, J. P. et al. (1998). *Principles of Development.* Current Biology, London.

Raff, R. A. (1996). *The Shape of Life: Genes, Development and the Evolution of the Animal Form.* University of Chicago Press, Chicago.

Willmer, P. (1990). *Invertebrate Relationships.* Cambridge University Press, Cambridge. The difficulties of constructing phylogeny from morphological evidence.

Moore, J. & Willmer, P. (1997). Convergent evolution in invertebrates. *Biological Reviews* **72**: 1–60.

Tautz, D. (1998). Debatable homologies. *Nature* **395**: 17–19.

Bolker, J. A. & Raff, R. A. (1996). Developmental genetics and traditional homology. *BioEssays* **18**: 489–494.

Conway Morris, S. (1993). The fossil record and the early evolution of the Metazoa. *Nature* **361**: 219–225.

Conway Morris, S. (1998). Eggs and embryos from the Cambrian. *BioEssays* **20**: 676–682.

Knoll, A. H. (1996). Breathing room for early animals. *Nature* **382**: 111–112.

Ramskold, L. & Hou, X. (1991). New early Cambrian animal and onychophoran affinities of enigmatic metazoans. *Nature* **351**: 225–228.

Wainright, P. O., Hinkle, G., Sogin, M. L. & Stickel, S. K. (1993). Monophyletic origins of the Metazoa: an evolutionary link with fungi. *Science* **260**: 340–342.

Kobayashi, M., Furuya, H. & Holland, P. W. H. (1999). Dicyemids are higher animals. *Nature* **401**: 762.

Ruiz-Trillo, I., Riutort, M., Littlewood, D. T. J. et al. (1999). Acoel flatworms: earliest extant bilaterian metazoans, not members of Platyhelminthes. *Science* **283**: 1919–1923.

Halanych, K. M., Bacheller, J. D., Aguinaldo, A. M. A. et al. (1995). Evidence from 18S ribosomal DNA that the lophophorates are protostome animals. *Science* **267**: 1641–1643.

De Robertis, E. M. & Sasai, Y. (1996). A common plan for dorsoventral patterning in Bilateria. *Nature* **380**: 37–40.

Slack, J. M. W., Holland, P. W. H. & Graham, C. F. (1993). The zootype and the phylotypic stage. *Nature* **361**: 490–492.

Aguinaldo, A. M. A., Turbeville, J. M., Linford, L. S. et al. (1997). Evidence for a clade of nematodes, arthropods and other moulting animals. *Nature* **387**: 489–493.

Paul, C. R. C. & Smith, A. B. (eds.) (1988). *Echinoderm Phylogeny and Evolutionary Biology.* Clarendon Press; Liverpool Geological Society, Oxford.

Wray, G. A. (1995). Punctuated evolution of embryos. *Science* **267**: 1115–1116.

Lowe, C. J. & Wray, G. A. (1997). Radical alterations in the roles of homeobox genes during echinoderm evolution. *Nature* **389**: 718–721.

Shear, W. A. (1992). End of the 'Uniramia' taxon. *Nature* **359**: 477–478.

Averof, M. & Akam, M. (1995). *Hox* genes and the diversification of insect and crustacean body plans. *Nature* **376**: 420–423.

Carroll, S. B. (1995). Homeotic genes and the evolution of arthropods and chordates. *Nature* **376**: 479–485

Fortey, R. A. & Thomas, R. H. (eds.) (1997). *Arthropod Relationships*. Chapman and Hall, London.

Akam, M. (1998). *Hox* genes: from master genes to micromanagers. *Current Biology* **8**: R676–678.

Glossary

Abdomen   The third and most posterior division of the body of an insect, applied also to describe crustaceans.

Abductor   When contracted, an abductor muscle separates two hard structures, or draws a limb away from the midline of the animal (contrast *Adductor*).

Aboral   Away from the mouth, for example the upper surface of a starfish or a sea urchin.

Actin   A globular protein which readily polymerises into a fibrous form and is an important component of *Microfilaments*. Actin constitutes the thinner fibrils involved in muscle contraction (Box 5.1).

Action potential   The nerve impulse passing along an *Axon*.

Active transport   The energy-requiring transport of a substance across a membrane against the concentration and/or the electrical gradient for that substance.

Adaptation   An adaptation is any change in structure or function that makes part or the whole of an organism better suited to its function or to its environment. It is the result of *Natural selection*.

Adaptive radiation   The divergent evolution of different species from a common ancestor, due to *Natural selection*.

Adductor   When contracted, an adductor muscle pulls together two hard structures (e.g. the two halves of a bivalve shell), or draws a limb towards the midline of the animal (contrast *Abductor*).

Alimentary canal   Food cavity (gut) which in most animals leads from the mouth to the anus.

Allele (short for *Allelomorph*)   Alternative form of *Gene* occupying the same locus (position) on a pair of *Chromosomes*.

Allelomorphic pair   See *Allele*.

Alternation of generations   Alternation between a *Haploid* and a *Diploid* generation within an organism's life cycle (does not apply to cnidarian *Polyps* and *Medusae*, where the *Chromosomes* are identical in number and kind in the two forms). Most animals confine the *Haploid* stage to the *Gametes*.

Ammonia (NH$_3$ ion NH$_4^+$)   The primary product of nitrogenous metabolism. Released directly into the environment by most marine organisms, but is very toxic; where water is not freely available, energy must be spent synthesising ammonia into less toxic forms (*Urea*, *Uric acid*, etc) for *Excretion*.

Amoebocyte   A cell that moves by *Amoeboid* movement, for example in a sponge.

Amoeboid   Any movement resembling that of the protistan, *Amoeba*, characterised by changes in shape of the moving cell.

Ampulla   A small vesicle containing liquid, for example those associated with the *Tube feet* in many echinoderms.

Anaerobic (or *Anoxic*)   Lacking oxygen. Anaerobic respiration occurs in many invertebrates: carbohydrates are incompletely broken down, producing lactic acid rather than carbon dioxide and water and making less energy available.

Antenna   Anterior sensory structure, typically a long, thin process in an arthropod, sensitive to mechanical and chemical stimuli.

Apodeme   A piece of the *Exoskeleton* of an arthropod, folded internally and attaching muscle.

Aquatic animal   One living in water.

Aristotle's Lantern   The feeding structure of a sea urchin.

Asexual reproduction   Usually by fission or budding, that does not involve either sexual fusion or *Meiotic* nuclear division.

Atrium   Chamber surrounding the *Pharynx* in some invertebrate chordates, receiving water through the pharyngeal gill slits and opening to the exterior by a more posterior pore.

Autotomy   The self-induced dropping off of part of an organism (often a limb) usually in response to threat from a *Predator*.

Axon   The main projection from a *Nerve cell*, along which nerve impulses pass.

Bilateral symmetry   When only one plane perpendicular to the oral/aboral axis can divide the animal into two mirror-image halves (contrast *Radial symmetry*).

Biramous   Having two branches.

Blastocoel   The primary cavity, appearing at the end of *Cleavage* in the developing embryo.

Blastomere   A cell within a *Blastula* or in early *Cleavage*.

Blastopore   The first opening of the embryonic gut. It becomes the mouth in *Protostomes* and the anus in *Deuterostomes*.

Blastula   Hollow ball of cells formed by the *Cleavage* of the developing egg.

Body cavity   A fluid-containing cavity between the gut and the outer surface of an animal, containing internal organs.

Botryoidal tissue   Tissue consisting of bunches of cells and containing a dark pigment, in leeches. Derived from *Parenchyma*, it invades the *Coelom*, leaving only a few channels serving for fluid transport.

Brain   A usually anterior aggregation of *Nerve cells* that acts as a sensory/motor exchange (in simple animals) or a controlling nerve centre for the whole body.

Buccal cavity   Space just inside the mouth, at the anterior end of the alimentary canal.

Buoyancy   Neutral buoyancy is achieved when a body floats at a constant level in the water. Buoyancy is said to be negative if the body is denser than the water and therefore tends to sink.

Byssus   Adhesive threads secreted by the foot of a bivalve mollusc, attaching the animal to the substratum.

Caecum (pl. *caeca*)   Pouch-like diverticulum of an animal's gut, usually functioning in digestion. In many herbivorous animals the caecum houses *Symbiotic* micro-organisms able to digest cellulose.

Carapace   The protective cuticular 'shell' of many crustaceans, overlying the main body *Cuticle*. Usually anterior only.

Catch muscle   Misleading name given to muscles such as bivalve *Adductors*, which can maintain low tension for a long time. Misleading because the muscle is not locked in position but requires a regular small nerve input and contraction is actively, if very economically, maintained.

Caudal   Pertaining to tail.

Cephalothorax   Combined head and thorax of many crustaceans.

Cercus (pl. *cerci*)   One of a pair of appendages on the last abdominal segment of an insect.

Chaeta (pl. *chaetae*)   Small stiff chitinous bristles of annelids and other worms (synon. *Setae*).

Chela (pl. *chelae*)   Claws, pincers.

Chelicera (pl. *chelicerae*)   One of the first pair of appendages on the *Prosoma* of a chelicerate. Often functions to seize and grasp food.

Chemosynthesis   Manufacture of organic food from its inorganic constituents using energy from chemical processes rather than (as in photosynthesis) light. Largely confined to bacteria, but many animals rely on *Symbiotic* chemosynthetic bacteria.

Chitin   Nitrogen-containing polysaccharide that is a major component of the *Cuticle* of many animals, notably arthropods. Chitin is soft, but combined with cross-linked protein chains it makes a hard and light covering.

Chlorocruorin   Rare iron-containing *Respiratory pigment* of some annelids.

Choanocyte   Characteristic cell constituting the inner layer of a sponge, consisting of a nucleated cell body in which a *Flagellum* is rooted and a collar made of *Microvilli* where food particles are ingested.

Chromatophores   Cells just under the skin of some animals, bearing pigment granules. Aggregation or dispersal of chromatophores changes the colour of the animal.

Chromosome   A thread-like structure in the cell nucleus carrying the *Genes*. Chromosomes are made of *DNA* and *protein*. They may be visible by light microscopy at various stages in cell division.

Cilium (pl. *cilia*)   A short (usually up to $10\,\mu$m long) hair-like thread borne on a cell, usually present in large numbers. Ciliary beat generates water currents that may move fluid through the animal or the animal through the water

Cladistic analysis aims to trace ancestry by constructing *Cladograms* based on combination of organisms sharing the same difference from the primitive condition, aiming to detect *Monophyletic groups*.

Cladogram A series of dichotomous branches depicting a hierarchy of recency of common ancestry, based on the presence of shared derived characters.

Cleavage The earliest divisions of the fertilised egg, proceeding by *DNA* replication and (usually) subdivision of the cytoplasm.

Clitellum The 'saddle' familiar in an earthworm and occurring in all oligochaetes and leeches. It collects the gametes and wriggles over the worm's head forming a cocoon where fertilisation and development occur.

Clone A group of genetically identical cells or organisms derived asexually from a single founder cell.

Cnidae See *Nematocysts*.

Coelenteron The water-containing inner cavity of cnidarians, open to the outside world. It is in the position of a gut and is not a *'Body cavity'* although it may serve as a hydrostatic *Skeleton*.

Coelom A *body cavity* entirely surrounded by *Mesoderm*: in section it appears between two layers of mesoderm.

Coelomoduct A *mesodermal* duct leading from the coelom to the outside world, as in the excretory and reproductive ducts of crustaceans, molluscs and many other animals (contrast *Nephridia*).

Collagen A fibrous *Protein* that forms a large part of animal connective tissue. Polypeptide chains of the protein form helical coils bound together to form fibrils which have great strength and little elasticity.

Commensal Sharing the food of another animal with which it lives. Often one member of the pair benefits and the other is not strongly affected.

Commitment (or *Determination*) The limitation of the developmental capacity which directs the cell towards a particular pathway, the prelude to *Differentiation*.

Compound eye A single eye formed of many thousands of optical units, *Ommatidia*, each with its own lens and light sensitive pigment. Occurs in arthropods and some other animals.

Contractile vacuole A structure within a cell which expands as it fills with water and contracts as it empties its contents to the exterior. A means of water regulation found in *amoebae* and freshwater sponges.

Convergent evolution Occurs when resemblances are due to common adaptations to a particular function or way of life, rather than due to close common ancestry. Cannot be distinguished satisfactorily from *Parallel evolution*.

Crossing over The exchange of parts of *Homologous chromosomes* when they come together at *Meiosis* (see Figure 1.1).

Cryptozoic animals 'Hiding animals' inhabiting dark secluded habitats such as crevices.

Ctenidium (pl. *Ctenidia*) A molluscan gill that is suspended in the *Mantle cavity*, used for respiration or trapping on small food particles. Primitively it consists of rows of filaments borne on a common axis.

Cuticle Outermost protective structure of an animal, often made partly of horny material such as *Collagen*, secreted by underlying cells. Flexibility may be combined with hard regions, as in the chitinous cuticle of arthropods, which forms an *Exoskeleton* needing to be moulted as the animal grows.

Cytoplasm   All the cell contents bound within an outer cell membrane, other than the *Nucleus*.

4d   A particular cell identifiable at a certain stage in *Spiral cleavage*. It normally gives rise to all the mesoderm in spirally cleaving animals.

Dendrite   Projection of *Nerve cell* bodies that make contact with other nerve cell projections.

Depolarisation   Applied to nerve: a decrease in the *Resting potential* across the nerve membrane.

Derived character   Secondary or specialised as opposed to *Primitive*.

Determination   See *Commitment*.

Detritus   Particle of organic matter in water, often decomposed, floating or associated with the substratum.

Deuterostome   The group of animals where the embryonic blastopore typically gives rise to the anus and the mouth develops secondarily elsewhere (e.g. echinoderms and chordates). Typically deuterostomes share other characters of early development (see Box 5.3).

Diapause   A period of arrested development and low oxygen consumption in an insect life cycle, whose onset is determined by some event in past time (often a change in day length).

Dichotomy   Division into two.

Differentiation   The appearance of differences between cells during development, preceded by *Commitment*.

Diffusion   Movement of a substance from a region of higher concentration to a region of lower concentration of that substance, freely into the surroundings or across a permeable or semi-permeable membrane. Diffusion is passive in that its occurrence does not require energy input.

Diploblastic   Having two cell layers only, the ectoderm and the endoderm, as in Cnidaria.

Diploid   Having the full number of *Chromosomes* normal for the species. The chromosomes are present in pairs, one being derived from each parent if reproduction is *Sexual*.

DNA (deoxyribonucleic acid)   The genetic material of all living organisms except some viruses. A major constituent of the *Chromosomes* in the *Nucleus* and also occurring in cytoplasmic organelles such as *Mitochondria*.

18 S rDNA   Gene coding for *18 S rRNA*.

Dominant   Genes expressed when present either as both members of an *Allelomorphic pair* or as only one member of that pair (contrast *Recessive*).

Dorsal   Upper surface or back of an animal.

Double helix   The genetic material: a spiral ladder-shaped molecule of *DNA* made by the cross-linking of pairs of bases in two chains of *Nucleotides*.

Ecdysis   The splitting and discarding of the old *Cuticle* which occurs when an arthropod moults.

Ecdysone   *Hormone* implicated in the moulting of arthropods and having various functions in other phyla.

Ecology   (Original scientific use) The interaction of organisms with each other and with the environment.

Ectoderm   Outermost cell layer of an animal's body: may form *Epidermis* (skin) and its outgrowths; *Nephridia*; nervous tissue.

Efficiency   (1) Physical efficiency is greatest for a structure or process requiring the minimum energy input for a given output. (2) Biological efficiency is greatest for a structure or process giving most selective advantage to an animal, which is frequently the same as (1) but not necessarily so.

Endocuticle   The soft innermost layer of arthropod *Cuticle*, containing *Chitin* not hardened by cross-linked *Proteins*. Reabsorbed into the body at moulting.

Endoderm   Innermost of the cell layers of a *Diploblastic* or *Triploblastic* body. Gives rise to the lining of the gut (alimentary canal) and its associated glands.

Endostyle   Ciliated mucus – secreting mid ventral groove in the *Pharynx* of some invertebrate chordates, able to take up iodine. Forerunner of the vertebrate thyroid gland.

Enterocoel   *Coelom* formed by outpouching from the embryonic gut, as in echinoderms and invertebrate chordates.

Epiboly   *Gastrulation* by the spreading of a sheet of outer cells around an inner layer, which hollows out to form the gut.

Epicuticle   The structure overlying the *Exocuticle* of arthropods, containing various amounts of wax.

Epidermis   The outermost living layer of many animals, characteristically a large number of similar cells in close contact.

Euryhaline   Aquatic animals able to tolerate a wide range of salinity, whether as *Osmoconformers* or *Osmoregulators*.

Evolution   Descent with modification, the process by which the diversity of organisms arose from the earliest forms of life. Evolution is the result of *Natural selection*: it has been called the modification of development by ecology.

Excretion   Removal of metabolic waste (i.e. waste materials formed by the body's own activity, not including undigested food). The term may be extended to all regulation of the internal body fluids, i.e. ion and water regulation, since the same organs are often involved.

Exocuticle   The layer of the arthropod *Exoskeleton* immediately under the *Epicuticle*, containing *Chitin* and hardened by cross-linked ('tanned') *Proteins*. Shed at moulting.

Exoskeleton   Hard *Skeleton* on the outer surface of an animal's body, such as the *Cuticle* of an arthropod.

Fecundity   Fertility: reproductive capacity.

Flagellum (pl. *flagella*)   Relatively long (150 μm) fine, whip-like projection from the surface of a cell, e.g. sperm, *Flame cell*, sponge *Choanocyte*. Like *Cilia*, the beat causes movement in the surrounding fluid; unlike cilia, flagella are long enough to have several waves at the same time.

Flame cell   Blind ending *Protonephridial* tube with a beating flagellar 'flame' drawing in water and solutes from the surroundings through the walls of the tube.

Fossils   Remnants (dating back to before the last Ice Age) of once-living animals, preserved in rocks or (less often) sediments, amber, ice, etc. Hard parts may be preserved, or footprints, tracks etc (see *Trace fossils*).

Free-living   (1) Not *Parasitic*. (2) Not confined to one spot but capable of locomotion in the environment.

Gamete   The *Haploid* unit (usually differentiated as egg or sperm) that fuses with another at *Sexual reproduction*.

Ganglion (pl. *ganglia*)   Collection of nerve cell bodies, as opposed to *Axons*.

Gastrodermis   Alternative description of an *Endodermal* layer, for example in Cnidaria.

Gastrulation   A series of cell movements in early development that brings the cells that will form internal organs to the inside of the developing embryo. Gastrulation converts a ball or sheet of cells into a multilayered structure within which the future germ layers take their correct relative positions.

Gel electrophoresis   Separation in an electric field of colloidal substances such as gels: a method of identifying the components of a mixture by the different rates at which they move between two applied electrodes.

Gene   Unit of heredity that passes from parent to offspring. Part of a molecule of *DNA* borne on a *Chromosome*, a gene consists of a sequence of nucleotide bases coding for all or part of a *Protein*. It is the shortest length of a DNA molecule that can undergo *Mutation* or *Recombination*.

Genetic drift   Random fluctuations in gene frequency that are not due to *Natural selection*, as when a small population within a species is isolated and *Evolution* depends on the genes that happen to be present.

Genotype   Genetic constitution of an organism, as opposed to the *Phenotype*.

Germ plasm   The part of an organism specialised for *Sexual reproduction*, i..e. for transmission of hereditary characters down the generations.

Glycogen   Large insoluble carbohydrate storage molecule: 'animal starch'.

Gnathobase   Process manipulating hard food, arising from an arthropod appendage.

Gonad   a structure producing *Gametes*, for example an ovary producing eggs and a testis producing sperm.

Gondwanaland   A former large land mass (some 135 million years ago) of the southern hemisphere from which South America, Africa, Madagascar, India, Australia and Antarctica are derived.

Guanine   Insoluble purine which is the end-point of nitrogenous metabolism in spiders. Can be excreted with very little loss of water.

Haemal   Concerning blood. Misapplied to one of the *Coeloms* of an echinoderm.

Haemerythrin   Rare iron-containing *Respiratory pigment* of some annelids.

Haemocoel   Blood-filled *Body cavity* derived from the *Blastocoel*, in arthropods and molluscs.

Haemocyanin   Copper-containing *Respiratory pigment* in many crustaceans and molluscs.

Haemoglobin   Iron-containing *Respiratory pigment* very widely occurring in almost every phylum.

Haltere   Modified hind wing or 'balancer' of a dipteran fly, in the form of a stick with a heavy head, beating with the wings. Strain receptors at the haltere base inform the fly of any irregularities needing to be corrected.

Haploid   Having half the number of *Chromosomes* normal for the particular species, as when *Gametes* are formed each with a single chromosome set.

Hermaphrodite   Having both male and female sex organs in one individual, which may be cross-fertilised, or more rarely self-fertilised.

Homeobox   Sequence of 180 *Nucleotide* bases that code for a sequence of amino acids, the *Homeodomain*. The homeobox has been strongly conserved during evolution: genes containing it regulate development. *Hox* genes are examples of homeobox-containing genes.

Homeodomain   Conserved sequence of 60 amino acid residues coded for by the *Homeobox*. Interaction between the homeodomain and target *DNA* sequences may determine whether or not the target genes are expressed.

Homologous   Morphological similarity due to a common evolutionary origin. This is a fundamental concept in tracing evolutionary relationships yet there is no simple, universally accepted, precise definition (e.g. as to whether a common developmental path should be invoked). See Chapter 20.

Homologous chromosomes   In this specialised usage 'homologous' approximates to 'similar'. Pairs of structurally similar *Chromosomes* can be identified at the outset of *Meiosis*; one member of each pair comes from each parent, and both members of a pair have the same pattern and positions of genes but the *Alleles* may be different.

Hormone   Substance made by a gland or (as often in invertebrates) a specialised *Nerve cell* (see *Neurosecretion*) and released into the blood, or into other tissues or fluids. The hormone travels to a distant site where it exerts its effect, often by regulating cell performance. Hormones in contrast to nerves may provide an alternative slower and longer-lasting method of coordinating the body.

Hox genes   Are *Homeobox*-containing highly conserved genes that determine the relative position of structures along the main body axis. They initiate a hierarchy of regulatory genes.

Hyperpolarisation   Increase in the *Resting potential* across a *Nerve cell* membrane.

Hypodermis   Layer of living cells underlying the thick outer *Cuticle* in a nematode. Equivalent to the *Epidermis* of other animals with cuticles.

Imago (adj. *imaginal*)   Adult insect.

Integument   Outer layer or skin (but see *Tegument*).

Intercellular   Between cells.

Interneuron   Nerve cell in a circuit between sensory input and motor output, which it may control by changes in *Polarisation* or *Depolarisation*; mostly interneurons do not give rise to action potentials. They are, for example, the sites of organisation of flight patterns in insects

Interstitial (1) Cnidarian unspecialised cells moving about between other cells and capable of differentiating into them. (2) Ecological microhabitat of spaces in between particles of soil or (most frequently) marine sand grains.

Intracellular Within a cell.

Introvert Front end of body of some worms which can be everted or retracted into the body.

Invagination Tucking in: applies to a common method of *Gastrulation*, by the infolding of a sheet of cells from a point on the *Blastula* surface.

Ionic regulation Selecting and keeping particular ions within the body at concentrations different from those of the surrounding water, even though osmotic regulation may not occur.

Isometric The sliding and cross-linking of *Myofilaments* in muscle contraction to build up tension with minimal shortening of the muscle as a whole (contrast *Isotonic*).

Isotonic (1) Muscle contraction by sliding of *Myofilaments* to shorten the muscle, with a minimum change in tension but considerable change in length (contrast *Isometric*). (2) The absence of net ion movement between two solutions separated by a semi-permeable membrane (often used where 'isoosmotic', equivalence of ion concentration, would be a more accurate description)

Junction potential Localised electrical changes under the endings of an arthropod motor nerve on the surface of a muscle, not normally propagated over the whole surface.

Labium The combined second maxillae (the most posterior pair of head appendages) of some insects and myriapods, forming a 'lower lip'.

Larva The pre-adult form in which some animals hatch from the egg (see Box 17.1 for the difficulties of fuller definition).

Linkage Describes the repeated absence of separation between parts of a *Chromosome* at *Crossing over* during nuclear division. The closer *Genes* are along a chromosome, the more likely they are to be linked: this was the original basis for all gene mapping.

Lophophore A circular or horseshoe-shaped fold of the body wall encircling the mouth (but not the anus), bearing hollow tentacles, each containing a branch of the *Mesocoel*. Cilia beat food-bearing water currents from the tip of each tentacle to the base, where the water leaves the animal.

Lung books (or *book lungs*) Respiratory organs of arachnids consisting of 'books' of blood containing leaflets with air channels in between.

Madreporite Echinoderm sieve plate connecting the water vascular system with the sea water outside.

Malpighian tubules Excretory organs in insects and some other arthropods. Long, narrow tubes growing out from the junction of the mid and hind gut, ending blindly in the blood-filled *Haemocoel*. Active ion transport from the blood to the tubule lumen draws other ions, water and excretory products into the tubule.

Mandibles Paired appendages just behind the mouth in many insects and crustaceans; hard structures used for seizing and biting food.

Mantle cavity Outside the main body wall of a mollusc, covered wholly or partly by the mantle (a fold of skin), the mantle cavity usually contains *Ctenidia* suspended in ambient water. In a pulmonate terrestrial snail it contains air instead of water.

Marine In the sea.

Maxilla (pl. *maxillae*) One of a pair of feeding appendages behind the mouth (in front of the *Mandibles*) in most insects and crustaceans.

Maxillipedes Thoracic appendages of crustaceans that assist the *Maxillae* in food collection.

Medusa (pl. *medusae*) The 'jellyfish' form of individual in Cnidaria, typically umbrella or bell shaped, freely floating or swimming, mouth downwards.

Meiosis (adj. *meiotic*) 'Reduction division' of the cell nucleus producing daughter cells with half the number of *Chromosomes* present in the parent.

Melting Point The temperature at which the solid and liquid phases of a substance are in equilibrium at a given pressure; usually refers to the temperature at which a solid melts at atmospheric pressure. Or, as here, the temperature at which double-stranded *DNA* dissociates to form single strands.

Mesocoel Coelomic cavity, the middle of the three divisions of the *Coelom* typically occurring in *Deuterostomes*. Provides the hydraulic system of a *Lophophore*.

Mesoderm Middle of the three cell layers of a *Triploblastic* animal. Gives rise to muscle, sex organs and many other structures.

Mesoglea Jelly-like material between the two cell layers of a cnidarian. containing cells and fibres derived from them. Much expanded in jellyfish.

Mesohyl Jelly-like substance between the outer cells and the *Choanocytes* in a sponge, containing wandering *Amoebocytes*.

Messenger RNA (mRNA) Molecules of RNA transcribed from the *DNA* of the genes and passing out of the nucleus into the cytoplasm and to the ribosomes. Characteristically short-lived except in egg cells, where long-lasting mRNA provides initial maternal instructions for development.

Metabolism The sum total of chemical reactions within the body of an organism.

Metabolite Product of *Metabolism*.

Metachronal rhythm Pattern of movement in which each unit (*Cilium* or muscle block) contracts just before or just after its neighbour.

Metameric segmentation Serial repetition of *Homologous* parts in an animal.

Metamorphosis Pronounced change in form during the animal life cycle.

Metanephridium *Nephridium* with an open end, occurring in some worm phyla. Combined with *Coelomoducts,* make the excretory organs of most annelids.

Microfilaments Contain the *Protein Actin* and are usually abundant under the cell membrane. Important in cell motility.

Microtubules Small tubular structures (15 to 25 nm) made of subunits of the *Protein* tubulin, occurring in large numbers freely in the cytoplasm and in cell organelles such as *Cilia* and *Flagella*. They appear to be important in transport of materials for cell motility.

Microvilli Small finger-like projections of the outer cell membrane, filled with

Mitochondria and occurring in large numbers at any site of active absorption, *Secretion* or sensory perception.

Mitochondrion (pl. *mitochondria*)   The organelle within the cytoplasm that bears the enzymes that carry out aerobic respiration. Very numerous in cells with high metabolic activity. Mitochondria contain their own *DNA*, different from that of the cell nucleus and in the circular form found in bacteria (a pointer to the *Symbiotic* origin of metazoan cells). Mitochondria are transmitted down the generations, normally only in the female line, and their DNA has become very important in tracing evolutionary relationships.

Mitosis   Division of the *Nucleus* producing two daughter nuclei with chromosomes identical in number and kind to those of the parent.

Molecular clock   Said to be regular if molecular evolution (e.g. substitutions in the *Nucleotide bases* of *DNA*) proceeds at a constant rate, so that the degree of difference between the DNA of two species can be used to measure time elapsed since their evolutionary divergence. The difficulties are that even if this form of *Neutral evolution* does occur, the rate of molecular change is known not to be constant in different species in different conditions.

Monophyletic   A group derived from a single ancestor, including that ancestor and all of its descendants. *Cladistic analysis* aims to reveal monophyletic groups.

Morphogen   Diffusing substance conferring the *Polarity* of an organism and/or initiating certain kinds of development that do not occur without it.

Mosaic   Made of many parts that together make a whole, applied: (1) to the development of *Spirally cleaving* eggs where *Commitment* occurs early; (2) to the *Compound eyes* of arthropods. Neither of these attributions is very accurate.

Mucus (adj. *mucous*)   Polysaccharide bound to a *Protein*, secreted by most animals in most phyla and possessing very important properties: it can be very liquid, gelatinous, or (with some chemical change) adhesive. It is used, for example, for lubrication, protection against desiccation, binding of food particles and adherence to the substratum.

Multicellular   Organisms (or parts of them) made of many cells (e.g. all Metazoa).

Mutation   Change in the structure of a *Gene* or *Chromosome* set, typically random in appearance. Mutations may lead to heritable variations, the raw material on which *Natural selection* acts, causing evolution.

Myofilaments   *Actin* and *Myosin* fibrils, which slide between each other when muscle contracts.

Myogenic   Originating in or produced by muscle cells, e.g. the rapid oscillation of the 'asynchronous' 'fibrillar' flight muscles of some small insects.

Myoglobin   Simple form of *Haemoglobin* present in various tissues; may store or facilitate the diffusion of oxygen.

Myosin   Protein with a globular head and a rod-like tail; the main constituent of muscle fibrils (see Box 5.1), constituting the thicker of the *Myofilaments*.

Natural selection   Process, discovered by Charles Darwin, that brings about evolution: the differential reproduction of individuals with attributes favourable to their particular habitat or way of life, imprecisely formulated as 'the survival of the fittest'. Results in the *Adaptation* of structure to function and organism to environment.

Nauplius Characteristically the first *Larva* of crustaceans, having simple eyes and three pairs of appendages.

Nematocysts (Cnidae) Stinging cells unique to the phylum Cnidaria. Reception of a stimulus and response (by the ejection of a stinging thread) is carried out by the single cell, which may act without nervous stimulation.

Neoteny *Paedomorphosis* by delay in the development of adult characters so that the animal reproduces while still in the larval form (cf. *Progenesis*).

Nephridium (pl. *nephridia*) Excretory organ made by intucking of an epidermal tube, consisting entirely of *Ectoderm*, occurring in several worm phyla, especially where there is great need for water regulation. May be a blind-ending *Protonephridium* (e.g. flame cell) or open-ended *Metanephridium*.

Nerve cell See *Neuron*.

Nerve net Network of nerve cells connected by non-polarised junctions and able to conduct in all directions. Occurs in various worms and in Cnidaria and Echinodermata where there is no controlling brain.

Neuron (neurone) Unit of conduction by the nervous system, consisting of a cell body with processes (*Dendrites*) and usually one long process, the *Axon*, conducting nerve impulses.

Neurosecretion Production of *Hormones* by specialised cells in the nervous system. Neurosecretory cells may be recognisable by large rounded nuclei in the cell body: the ends of the axons secrete hormones acting at a distance rather than nerve transmitters that bridge a *Synapse*.

Neutral evolution Descent with modification not due to the action of *Natural selection*, causing accumulation in the genome of selectively neutral *Mutations*. The extent to which it occurs continues to be controversial.

Niche An ecological niche defines not only where an organism lives but also its type of food and general requirements.

Notochord Dorsal stiffening rod characteristic of Chordata, made of closely fitting cells with hydrostatic pressure due to intracellular *Vacuoles* and covered in a fibrous sheath.

Nucleolus Region of the cell nucleus especially rich in *RNA* and *Protein*. The site of origin of *Ribosomes*.

Nucleotide Unit of structure of nucleic acids, *DNA* and *RNA*, consisting of a sugar (deoxyribose or ribose) with one hydroxyl (–OH) group bound to a purine or pyramidine nitrogenous base.

Nucleus Cell organelle containing the *Chromosomes*, hence the *Genes*, and bounded by a nuclear membrane. Present in all cells, at their origin at least.

Numerical taxonomy See *Phenetic taxonomy*.

Nymph Old-fashioned term for the larva of an exopterygote insect.

Oesophagus Part of the gut between the *Buccal cavity* and the stomach: there may or may not be a distinct *Pharynx* between the buccal cavity and the oesophagus.

Ommatidium One of the light-sensitive units in an arthropod *Compound eye*, consisting of a cuticular lens and a number of light-detecting *Retinula* cells, the whole being surrounded by pigment.

Ontogeny Development of an individual from an egg to an adult.

Opisthosoma Posterior division of the body of a chelicerate, without locomotor appendages but with gills in aquatic forms.

Osculum (pl. *oscula*) Large hole in a sponge through which a jet of water is ejected.

Osmoconformer Animal (such as most marine invertebrates) whose internal body fluids have approximately the same osmotic potential as the surrounding water.

Osmole The osmotic effect of a solute when the molecular weight of the solute is dissolved in a litre of water. For sea water the total osmotic effect of the solutes is 1 osmole (1000 milliosmoles, mosm.).

Osmoregulation Maintenance of the internal body fluids at a concentration (and hence osmotic potential) different from that of the surrounding water.

Ossicle Skeletal unit of an echinoderm: a calcareous plate separated by living tissue from its neighbours, at least while it develops.

Ostium (pl. *ostia*) (1) one of many small holes perforating the body of a sponge, through which water enters; (2) holes in an arthropod heart through which blood enters from the *Haemocoel*.

Outgroup Group of species used to define the root of a *Cladogram*, chosen as evolutionarily close to the group in the cladogram but not a part of it.

Ovoviviparity Embryos are retained in yolky eggs within the mother's body while they develop, but are not fed by the mother (contrast *Viviparity*).

Oxygen tension Measure (by pressure, expressed as millimetres of mercury (mmHg, $1\,mmHg \approx 133.3\,Pa$) in a barometer) of the availability of oxygen in a solution.

Paedomorphosis Development to reproductive maturity of the juvenile form of an animal, either by *Neoteny* or by *Progenesis*.

Papilla A cone-shaped projection from the surface of an organ or an organism.

Papula Respiratory projections from the echinoderm water vascular system.

Parallel evolution See *Convergent evolution*.

Paraphyletic Assembly including some but not all members of a *Monophyletic* group.

Parapodium (pl. *parapodia*) Paddle-like lobes projecting from the sides of the segments of polychaete worms. Used mainly for moving water, sometimes involving locomotion.

Parasite Specialised form of *Predator* that does not kill its prey: it lives inside, or attached to the outer surface of, a host (of another species) and feeds on the host's living body. The host derives no benefit from the association.

Parenchyma Unspecialised middle layer of some simple *Triploblastic* animals. Characterised by few cells, assorted fibres and intercellular spaces filled with liquid.

Parsimony Originally defined in the application of 'Occam's Razor' (the acceptance of simple explanations in preference over more complicated ones). Parsimony has been annexed by cladists to describe the choice of one of many possible *Cladograms* postulating the smallest number of evolutionary steps, i.e. the minimum need to postulate convergence. This process minimises the perception of convergence and may or may not be correct.

Parthenogenesis Development from an unfertilised egg in a female animal.

Pedicellaria Small, stalked pincers on the surface of an echinoderm, used for grasping and defence.

Pedipalp One of the second pair of appendages on the *Prosoma* of a chelicerate. Functions vary from seizing and killing prey to use as walking or digging legs.

Pelagic Organisms swimming or drifting in the main body of the sea or a lake, whether *Planktonic* or in deeper waters.

Pericardium Membrane round the heart. The pericardial cavity is a *Coelom* containing the heart.

Peristalsis Movement of part or the whole of an organism by waves of alternate contraction and extension, for example movement of food through the intestine or of an earthworm through soil, achieved by circular and longitudinal muscles contracting in turn.

Pharynx (adj. *pharyngeal*) The part of the alimentary canal behind the buccal cavity and anterior to the oesophagus.

Phenetic taxonomy (numerical taxonomy) proceeds by putting together those organisms with the greatest number of common characters.

Phenotype Manifested attributes of an organism. Due primarily to *Gene* activity, but contrast *Genotype*.

Pheromone *Hormone* acting outside the body.

Phylogenetic The relationship between organisms and groups of organisms that is due to their evolutionary history. Description of taxonomy based on *Cladistics*.

Phylogeny The pattern of evolutionary relationships.

Phylotypic stage The stage when *Gastrulation* is complete but before cell *Differentiation*. At this stage the members of a phylum show maximum similarity to each other, i.e. the characters of the body plan are most clearly revealed.

Phylum (pl. *phyla*) Group of animals with a common body plan.

Pinacocyte Outer cell of a sponge.

Plankton Surface waters of the sea or a freshwater lake. Applied to small animals and plants drifting or swimming in surface waters; they provide a very important source of food.

Pluteus Echinoderm larva characteristic of echinoids and ophiuroids.

Polarisation (of a nerve membrane) Presence of the *Resting potential*.

Polarity (1) Definition of the axis of an organism or part – up and down, front and back; (2) direction or order of succession in time or space.

Polymorphism Different forms occurring within one *Species*.

Polyp One form of individual within the phylum Cnidaria, typically fixed at one end and bearing tentacles round the mouth at the other end.

Polyphyletic Assemblage of animals with different ancestral lineages.

Predator Animal that kills and feeds on another animal.

Primitive Earliest, not necessarily simplest, example of a character or organism; may or may not be preserved to the present time.

Proboscis (pl. *proboscides*) Any tubular projection at the anterior end of an animal, whether it is part of the gut or separate from it (as characterises nemertines), or is the trunk of an elephant.

Progenesis *Paedomorphosis* by acceleration of development of the gonads; the animal reproduces in the larval form and the generation time is shortened.

Proprioceptor Sensory receptor responding to changes within the body, for example stretch receptors in muscle.

Prosoma Anterior division of the body in chelicerates.

Protandrous *Hermaphrodite* in which the male organs develop and function first.

Protein Large group of complex nitrogen-containing organic compounds, made up of polypeptides consisting of strings of amino acids. The primary constituent of living matter, including enzymes.

Protogynous *Hermaphrodites* in which the female organs develop and function first.

Protonephridium *Nephridium* with a closed end, as in *Flame cells*. Occurs in platyhelminths, nemertines, a few polychaetes and larvae.

Protostome *Triploblastic* animals in which the *Blastopore* usually becomes the mouth, with other distinctive features of early development (see *Deuterostome*).

Pseudocoel *Body cavity* which is not a true *Coelom* but a persistent *Blastocoel* or (more rarely) derived from cell *Vacuoles*.

Pupa Stage between larva and adult in an endopterygote insect. Known as a resting stage, but is a period of great internal reorganisation as larval tissues are resorbed and the adult rudiment develops.

RNA (ribonucleic acid) The nucleic acid made from *DNA* and concerned with *Protein* synthesis. RNA is itself the genetic material in some viruses, and perhaps was so at the origin of life. Differs from DNA in that the four bases are adenine, cytosine, and guanine, with uracil in place of thymidine. Various forms of RNA occur in all cells, with different functions from transcription of the genetic code and its translation at the *Ribosomes*.

18 S rRNA The small subunit of ribosomal *RNA* (18S describes the centrifugation coefficient obtained in isolating that subunit) often used in molecular taxonomy.

Radial cleavage *Cleavage* in which the axis of cell division is directly perpendicular to the mother cell, so that cells lie directly beside or on top of one another (contrast *Spiral cleavage*).

Radial symmetry Symmetry about any plane perpendicular to the oral/aboral axis.

Radula Rasping tongue located in the *Buccal cavity* of a mollusc: unique and diagnostic feature of the phylum. Most commonly a chitinous strip with teeth constantly renewed, but may become a drill or piercing organ.

Recessive Genes expressed only when present as both members of an *Allelomorphic pair*; masked when the other member of the pair is different and *Dominant*.

Recombination Rearrangement of genes occurring at *Crossing over* in *Meiosis*.

Regulative In embryological development, applied to late *Commitment* of cells, as in most *Radially cleaving* eggs.

Respiratory pigment Protein molecule, pigmented and containing metal, that combines reversibly with oxygen and therefore can transport or store it.

Resting potential Small potential difference across a nerve membrane, the outside being positively charged, due to outward movement of potassium ions.

Retractor Muscle, contraction of which withdraws a structure back into the body.

Rhabdome Closely packed *Microvilli* containing the visual pigment, e.g. in the central region of the *Retinula* cells of an arthropod *Ommatidium*.

Rhynchocoel Fluid-filled cavity, lined with mesoderm, housing the *Proboscis* of a nemertine worm.

Ribosomes Granular bodies in the cytoplasm of all cells, containing *RNA* of several kinds. Sites of assembly of amino acids, translation of the genetic code and protein synthesis.

Rudimentary Structure at an early stage of its evolution, recognisable by comparison with its appearance later in evolution (contrast *Vestigial*).

Schizocoel *Coelom* formed by splitting of embryonic mesoderm, as in most *Protostomes* (contrast *Enterocoel*).

Secretion Cellular manufacture of useful substances and the act of discharging these from the cell.

Sedentary Attached to a substratum but able to move away from the place of attachment (contrast *Sessile*).

Segmentation Subdivision of an animal into a number of equivalent parts, contained between the anterior and posterior end of an organism (*Metameric segmentation*).

Sensillum (pl. *sensilla*) Hairs, bristles or pads on the arthropod *Skeleton* able to be sensitive to mechanical or chemical changes in the environment.

Septum (pl. *septa*) Internal divisions within an animal (e.g. between the segments of an annelid worm).

Sequencing A *Gene*: identifying all the nucleotides in that part of a *DNA* molecule that constitutes a gene and finding their relative positions.

Sessile Organisms permanently attached to a substratum (contrast *Sedentary*).

Seta See *Chaeta*.

Sexual reproduction Involving the fusion of two *Gametes*, each containing half the number of *Chromosomes* characteristic of the species.

Shrimp Strictly applies to one group of decapod crustaceans but applied more widely to any crustaceans filter-feeding with their swimming appendages.

Skeleton Structure providing mechanical support for the body and (usually) enabling muscles to be re-extended after contracting. May be hard or hydrostatic.

Soma The part of a body that is not specialised for *Sexual reproduction*.

Species (singular species, abbrev. 'sp': plural also species, abbrev. spp.) Population of interbreeding organisms that normally produce fertile young. A species is designated binomially, by a generic followed by a specific name.

Spicule   Small spine or spike, for example the hard structures in a sponge.

Spiracle   Opening at the body surface, for example of a part of the *Tracheal* apparatus of a terrestrial arthropod.

Spiral cleavage   Method of *Cleavage* where the axis of cell division at early stages is at an oblique angle to the mother cell, so that the cells form an upward spiral as they divide (contrast *Radial cleavage*).

Stenohaline   Aquatic animals restricted to a narrow range of salinity.

Striated muscle   Fast contracting muscle with the *Actin* and *Myosin* filaments arranged to give a striped appearance under a light microscope (see Box 5.1). Occurs in nearly all animal phyla.

Stylet   Hard, pointed structure able to penetrate the tissues of other animals.

Success   Of a living form is defined by its survival. The concept may include long persistence down the ages, large numbers of individuals (and of species, i.e. diversity), but not complexity nor 'progress'.

Supra-oesophageal ganglion   See *Brain*.

Suspension feeding   Capture of food particles floating in water.

Symbiosis   Close association of two organisms of different species that confers mutual benefit.

Synapomorphy   Presence of shared derived characters.

Synapse   The gap between two nerve cells or a nerve and a muscle or gland. Conduction across a synapse is mediated either chemically or electrically and occurs in one direction only.

Syncitium   Group of cells without cell boundaries between the nuclei, whose division has not been accompanied by cytoplasmic division.

Taxonomy   Systematic ordering of the products of classification of organisms.

Tegument   Particular type of outer layer in place of the epidermis in parasitic platyhelminths.

Tissue   An association of similar cells, usually connected by intercellular material. Organs consist of a number of coordinated tissues.

Torsion   Twisting of the body during development through 180°, as is characteristic of gastropod molluscs. Torsion brings the *Mantle cavity* from the back to the front of the animal, makes the gut U-shaped and may twist the nerves into a figure of 8.

Trace fossils   Imprints on the environment left by animal activity of long ago in the form of trails, burrows or impressions of soft tissues.

Trachea (pl. *tracheae*)   Air-filled cuticular tube, part of a system conducting the atmospheric environment to the tissues of a terrestrial arthropod (to every cell of the body, in insects).

Tracheole   Fine terminal tip of an insect *Trachea* penetrating between or even into cells. Cuticular, but not shed at moulting.

Trichotomy   Division into three.

Triploblastic   Having three cell layers, *Ectoderm*, *Mesoderm* and *Endoderm*, as in nearly all animals.

Trochophore   Early planktonic larva with two ciliary bands in front of the mouth and ciliary tufts elsewhere, as in annelids and molluscs.

Tube foot   Thin-walled epidermal projection of the coelomic water vascular system in echinoderms. Unique to the phylum, tube feet are sensory and respiratory and often involved in feeding and locomotion.

Ultrafiltration   Filtration on a small scale. Pressure (in most excretory organs) or suction (*Flame cells*) forces water and some solutes through a semi-permeable membrane into a tubule.

Ultrastructure   Level of structure that is almost molecular, being too small to be seen by light microscopy, but capable of being studied by electron microscopy.

Uniramous   Unbranched (i.e. having a single branch only).

Unsegmented   See *segmentation*.

Urea $CO(NH_2)_2$, readily made by condensing two molecules of ammonia with one of carbon dioxide, is the main nitrogenous excretory end-point of many aquatic animals.

Uric acid   Insoluble end-point of nitrogen metabolism in many terrestrial animals, since it can be stored dry and excreted with very little water loss.

Vacuole   Small membrane- bound intracellular sac containing fluid. May contain food, enzymes or water voided at intervals (*Contractile vacuoles*).

Variation   Difference between individuals within a *Species*.

Vegetal   Yolk-containing region of an egg or *Blastula* at the opposite end from the 'animal' pole where the nucleus is situated.

Ventral   Lower or front surface, 'tummy'.

Vestigial   Part of an organism that has diminished during its evolutionary history but not quite disappeared (contrast *Rudimentary*).

Viviparity   Developing young are housed in the mother's living body and fed from her own resources.

Water vascular system   Transport system in which the transporting fluid is water, not blood, as in sea anemones and the *Coelom* bearing the *Tube feet* in echinoderms.

Zooid   Individual within a colony; applied to bryozoans and a few other animals.

Zootype   Defined by the set of regulatory *Genes*, including *Hox* genes, which are apparently found in all animals and only in animals.

Zygote   Product of fusion of *Gametes* in *Sexual reproduction*, it is the single *Diploid* cell from which an organism develops.

Index

Italic numbers denote pages with illustrations; bold numbers denote major accounts of groups or topics. Box numbers are given in brackets after the relevant page numbers.